T0296126

Cambridge Tracts in Mathematics and Mathematical Physics

No. 13

The Twenty-Seven Lines upon the Cubic Surface

THE TWENTY-SEVEN LINES
UPON
THE CUBIC SURFACE

by

ARCHIBALD HENDERSON, M.A., Ph.D.

Professor of Pure Mathematics, University of North Carolina, U.S.A.

Cambridge:

at the University Press

1911

CAMBRIDGE
UNIVERSITY PRESS

University Printing House, Cambridge CB2 8BS, United Kingdom

Cambridge University Press is part of the University of Cambridge.

It furthers the University's mission by disseminating knowledge in the pursuit of education, learning and research at the highest international levels of excellence.

www.cambridge.org
Information on this title: www.cambridge.org/9781107493513

First published 1911
First paperback edition 2015

A catalogue record for this publication is available from the British Library

ISBN 978-1-107-49351-3 Paperback

NOTE

I TAKE pleasure in expressing my thanks to those who have read the present memoir in manuscript: Professor E. H. Moore and Dr L. E. Dickson of the University of Chicago, and Dr H. F. Baker of Cambridge University. To their kindly suggestions are due many improvements in the text. For any faults or errors, I alone am responsible.

A. H.

LONDON.
July, 1911.

CONTENTS

HISTORICAL SUMMARY

While it is doubtless true that the classification of cubic surfaces is complete, the number of papers dealing with these surfaces which continue to appear from year to year furnish abundant proof of the fact that they still possess much the same fascination as they did in the days of the discovery of the twenty-seven lines upon the cubic surface. The literature of the subject is very extensive. In a bibliography on curves and surfaces compiled by J. E. Hill, of Columbia University, New York, the section on cubic surfaces contained two hundred and five titles*. The Royal Society of London Catalogue of Scientific Papers, 1800–1900, volume for *Pure Mathematics* (1908), contains very many more.

The first paper that deals specifically with the cubic surface was by L. Mossbrugger, "Untersuchungen über die geometrische Bedeutung der constanten Coefficienten in den allgemeinen Gleichungen der Flächen des zweïten und dritten Grades," which appeared in the first volume of the *Archiv der Mathematik und Physik*, 1841.

The theory of straight lines upon a cubic surface was first studied in a correspondence between the British mathematicians Salmon and Cayley; and the results were published, *Camb. and Dublin Math. Journal*, Vol. IV. (1849), pp. 118–132 (Cayley), pp. 252–260 (Salmon). The observation that a definite number of straight lines must lie on the surface is initially due to Cayley, whereas the determination of that number was first made by Salmon†.

The basis for a purely geometric theory of cubic surfaces was laid by Steiner‡ in a short but extremely fruitful and suggestive memoir. This paper contained many theorems, given either wholly without

* *Bull. Am. Math. Soc.* Vol. III. (1897), pp. 136–146.

† Salmon, *Geometry of Three Dimensions*, 4th edition, § 530, note. Cf. also Cayley, *Coll. Math. Papers*, Vol. I. note, p. 589.

‡ "Ueber die Flächen dritten Grades," read to the Berlin Academy, 31st January, 1856; *Crelle's Journ.*, Vol. LIII.

proof, or with at most the barest indication of the method of derivation—a habit of "*ce célèbre sphinx*," as he has been styled by Cremona.

On account of what Cayley described as the "complicated and many-sided symmetry" among the relations between the twenty-seven lines upon the cubic surface, great difficulty was at first experienced in obtaining any adequate conception of the complete configuration. The notation first given by Cayley was obtained by starting from some arrangement that was not unique, but one of a system of several like arrangements; but it was so complicated as scarcely to be considered as at all putting in evidence the relations of the lines and triple tangent planes. Hart gave a very elegant and symmetrical notation for the lines and planes, an account of which is to be found in the original paper by Salmon*, who also gave a notation of limited usefulness. Schläfli† it was who invented the notation which may be called epoch-making—that of the double six‡; and this notation has remained unimproved upon up to the present time. This notation is one out of a possible thirty-six of like character among the twenty-seven lines. More recently, Taylor§ has devised a notation for the lines independent of any particular initial choice; but this cannot be regarded as an improvement upon the notation devised by Schläfli.

The foundations for subsequent analytic investigations concerning the twenty-seven lines were laid, as shown, by Cayley and Salmon. Indeed Sylvester‖ once remarked, in his characteristically florid style: "Surely with as good reason as had Archimedes to have the cylinder, cone and sphere engraved on his tombstone might our distinguished countrymen leave testamentary directions for the cubic eikosiheptagram to be engraved on theirs."

The first significant papers on cubic surfaces from the synthetic standpoint, following Steiner's memoir above mentioned, were by Cremona and Rudolf Sturm. These were two of the four papers submitted in competition for the prize offered by Steiner through the Royal Academy of Sciences of Berlin in 1864, which was divided between Cremona and Sturm on Leibniz Day, 1866. The beauty and simplicity of many of the methods employed in these papers eminently justified Steiner's original remark: "Es ist daraus zu sehen, dass diese

* *Infra*, § 4.

† *Quart. Journ.* Vol. II. (1858), pp. 55–65, 110–120.

‡ For the history of the double six theorem, see *infra*, § 6.

§ *Philos. Trans. Royal Soc.* Vol. CLXXXV. (1894), Part I. (A), pp. 37–69.

‖ *Proc. London Math. Soc.* Vol. II. p. 155.

Flächen fortan fast eben so leicht und einlässlich zu behandeln sind, als bisher die Fläche zweiten Grades." Cremona's "Mémoire de géométrie pure sur les surfaces du troisième ordre" is found in *Crelle's Journal**. Sturm's memoir was subsequently expanded into a treatise†.

Schläfli (l.c.) first considered a division of the general surface of the third order into species, in regard to the reality of the twenty-seven lines. He contented himself with a mere survey of the problem. This was in 1858. In 1862, F. August‡ published a more elaborate investigation of the subject. In 1863 appeared a valuable memoir by Schläfli§ treating the subject in great detail. He makes there, as the title of the paper indicates, a division of the surface into types—depending upon the nature of the singularities. His classification was adopted by Cayley in his "Memoir on Cubic Surfaces"‖.

If Cayley and Salmon had wished to follow Sylvester's advice and to insert a clause in their wills, directing that a figure of the eikosiheptagram be engraved upon their monuments, they would have had no certainty of the correct fulfilment of their directions until the year 1869, when Christian Wiener made a model of a cubic surface showing twenty-seven real lines lying upon it¶. This achievement of Wiener's, Sylvester** once remarked, is one of the discoveries "which must for ever make 1869 stand out in the Fasti of Science." In his address to the Mathematical and Physical Section of the British Association, H. J. S. Smith†† states that "a model showing the distribution in space of the lines themselves, unaccompanied by the surface on which they lie, has been constructed by Professor Henrici"; but Henrici does not seem to have published any paper on the subject.

In 1872, Clebsch and Klein, at Göttingen University, considered the question as to the shape of surfaces of the third order. Clebsch

* Vol. LXVIII. (1868), pp. 1–133.

† *Synthetische Untersuchungen über Flächen dritter Ordnung*, B. G. Teubner, Leipzig, 1867.

‡ *Disquisitiones de superficiebus tertii ordinis*, Dissert. inaug. Berolini, 1862.

§ "On the Distribution of Surfaces of the Third Order into Species, in reference to the presence or absence of Singular Points and the reality of their Lines," *Philos. Trans. Royal Soc.* Vol. CLIII. (1863), pp. 193–241.

‖ *Philos. Trans. Royal Soc.* Vol. CLIX. (1869), pp. 231–326.

¶ Cf. Cayley, *Trans. Camb. Philos. Soc.* Vol. XII. Part I. (1873), pp. 366–383, where a description of the model is given.

** *Proc. London Math. Soc.* Vol. II. p. 155.

†† "Geometrical Instruments and Models," *South Kensington Museum Handbook to the Special Loan Collection of Scientific Apparatus* (1876), pp. 34–54.

first constructed a model of the *diagonal surface* with twenty-seven real lines. "Instigated by this investigation of Clebsch," says Klein, "I turned to the general problem of determining all possible forms of cubic surfaces. I established the fact that by the principle of continuity all forms of real surfaces of the third order can be derived from the particular surface having four conical points*." Klein's method established completeness of enumeration—the consideration of fundamental importance. Klein exhibited a complete set of models of cubic surfaces at the World's Exposition in Chicago in 1894, including Clebsch's symmetrical model of the diagonal surface and Klein's model of the cubic surface having four real conical points. Models of the typical cases of all the principal forms of cubic surfaces have been constructed by Rodenberg† for Brill's collection; and these plaster models may now be purchased. Blythe has constructed models of certain types of cubic surfaces, and illustrated in some detail the character of the changes that take place under certain conditions‡. The list of those who have written on the mechanical construction of the configurations of the lines upon a cubic surface and the general collocation of the lines upon the surface includes the names of Cayley, Frost, Zeuthen, De Vries, Taylor and Blythe§.

The configuration of the twenty-seven lines is not only of great interest *per se*, but also because of its close association with, and relation to, other remarkable configurations. It was also in the year 1869—the year over which Sylvester waxed dithyrambic—that Geiser‖ showed that the projection of a cubic surface from a point upon it, on a plane of projection parallel to the tangent plane at that point, is a quartic curve; and that every quartic curve can be generated in this way. He showed the mutual interdependence of the configurations of the twenty-eight bitangents to a plane quartic curve and the twenty-seven lines upon a cubic surface, and the method of derivation of either configuration from the other. By making use of

* *Lectures on Mathematics*, Evanston Colloquium, 1894, Macmillan and Co. Cf. Klein's paper, "Ueber Flächen dritter Ordnung," *Math. Ann.* Bd. VI. (1873), pp. 551–581, where are to be found figures and sketches of surfaces having one conic node, symmetrical in form.

† "Zur Classification der Flächen dritter Ordnung," *Math. Ann.* Bd. XIV. pp. 46–110.

‡ *On Models of Cubic Surfaces.* Cambridge University Press, 1905.

§ Cf. *infra*, §§ 18, 21.

‖ *Math. Ann.* Bd. I. (1869), pp. 129–138. Cf. also *Crelle's Journ.* Vol. LXIII. p. 377.

Geiser's results, Zeuthen* obtained a new demonstration of the theorems of Schläfli† upon the reality of the straight lines and triple tangent planes of a cubic surface. He proves the reality of all the twenty-eight bitangents to the quartic in the case when the curve consists of four separate closed portions. In the next year, he points out the important connection between Klein's researches on cubic surfaces (l.c.) and his own researches on plane quartic curves. If a surface with four conic nodes be chosen, the resulting quartic has four double points. By the principle of continuity, the four ovals of the quartic are readily obtained; and this, as Zeuthen showed, corresponds to Klein's derivation of the diagonal surface from the cubic surface having four conic nodes‡. Timerding has shown that it is feasible to derive the properties of the plane quartic curve and its bitangents from the known properties of the cubic surface and its straight lines, and *vice versa*§.

In 1877, Cremona‖ was first to show that the Pascalian configuration might be derived from the configuration of the twenty-one lines upon the surface of the third degree with one conical point (Species II in Cayley's enumeration) by projection from the conical point. Mention should also be made here of the elaborate paper of Bertini¶.

Among recent investigations on the theory of the cubic surface, the allied problems of the twenty-seven lines, and the bitangents to the plane quartic curve, with generalizations to higher dimensions, are the papers, here given in chronological order, of: Richmond (*Camb. Phil. Proc.* Vol. XIV. 1908, pp. 475–477), Dixon (*Quart. Journ.* Vol. XL. 1909, pp. 381–384; *ibid.* Vol. XLI. 1910, pp. 203–209), Burnside (*Camb. Phil. Proc.* Vol. XV. 1910, pp. 428–430), Miss M. Long (*Proc. London Math. Soc.* Ser. 2, Vol. IX. 1910, pp. 205–230), Baker (*Proc. London Math. Soc.* Ser. 2, Vol. IX. 1910, pp. 145–199; *Proc. Royal Soc.* A, Vol. LXXXIV. 1911, pp. 597–602), and Bennett (*Proc. London Math. Soc.* Ser. 2, Vol. IX. 1911, pp. 336–351). In the first of his two papers above mentioned, Baker gives a proof of the theorem

* *Math. Ann.* Bd. VII. (1874), pp. 410–432.

† *Quart. Journ.* Vol. II. (1858); *Philos. Trans. Royal Soc.* Vol. CLIII. (1863).

‡ *Math. Ann.* Bd. VIII. (1875), pp. 1–30.

§ *Crelle's Journ.* Vol. CXXII. (1900), pp. 209–226.

‖ *Reale Accademia dei Lincei*, Anno CCLXXIV. (1876–77), Rome. Also cf. *infra*, §§ 45–6.

¶ "Contribuzione alla teoria delle 27 rette e dei 45 piani tritangenti di una superficie di 3° ordine," *Annali di Matematica* (1883–4), II. 12, pp. 301–346.

that there are only two ways in which a Steiner system of bitangents arises from lines of the cubic surface, and the new theorem that the six transversals from any point of the cubic surface, one to each of the opposite line pairs of a double six, lie on a quadric cone.

The theory of *varieties* of the third order, that is to say, curved geometric forms of three dimensions contained in a space of four dimensions, has been the subject of a profound memoir by Corrado Segre*. The depth and fecundity of this paper is evinced by the fact that a large proportion of the propositions upon the plane quartic and its bitangents, Pascal's theorem, the cubic surface and its twenty-seven lines, Kummer's surface and its configuration of sixteen singular points and planes, and on the connection between these figures, are derivable from propositions relating to Segre's cubic variety, and the figure of six points or spaces from which it springs. Other investigators into the properties of this beautiful and important locus in space of four dimensions and some of its consequences are Castelnuovo and Richmond†.

The problem of the twenty-seven lines is full of interest from the group theoretic standpoint. In 1869, Camille Jordan‡ first proved that the group of the problem of the trisection of hyperelliptic functions of the first order is isomorphic with the group of the equation of the twenty-seventh degree, on which the twenty-seven lines of the general surface of the third degree depend. In 1887, Klein§ sketched the effective reduction of the one problem to the other. In 1887–1890, Maschke in a series of papers set up the complete form-system of a quaternary group of 51840 substitutions‖. In 1893, Burkhardt¶, on the basis of Klein's paper above mentioned, these papers of Maschke, and a paper by Witting**, carried out the work sketched by Klein—the reduction of the one problem to the other.

* *Atti d. R. Accad. di Scienze di Torino*, Vol. XXII. (1887), pp. 547–557. Cf. also *Memorie d. R. Accad. di Scienze di Torino*, Series 2, Vol. XXXIX. (1889), pp. 3-48.

† Cf. Richmond's paper, *Quart. Journ.* Vol. XXXIV. No. 2 (1902), pp. 117–154 for references.

‡ *Comptes Rendus*, Vol. LXVIII. (1869), p. 865 *et seq.* Cf. also *Traité des Substitutions*, p. 216 *et seq.*, p. 365 *et seq.*

§ Extrait d'une lettre adressée à M. C. Jordan, *Journ. de Liouville*, Series 4, Tome IV. (1888), p. 169 *et seq.*

‖ *Math. Ann.* Bd. XXX. (1887), pp. 496–515; *Gött. Nach.* (1888), pp. 76–86; *Math. Ann.* Bd. XXXIII. (1889), pp. 317–344; *Math. Ann.* Bd. XXXVI. (1890), pp. 190–215.

¶ *Math. Ann.* Bd. XLI. (1893), pp. 309–343.

** *Math. Ann.* Bd. XXIX. (1887).

Since Jordan's paper appeared in 1869, many writers have studied the Galois group of the equation of the twenty-seven lines. Dickson* has led in this investigation, publishing a number of papers on the subject. Among those who have treated the problem of the twenty-seven lines as a problem in substitution groups or Galoisian groups may be mentioned Kühnen†, Weber‡, Pascal§, and Kasner||. This last paper is in close contact with the investigations of Moore¶ and Slaught** on the cross-ratio group of Cremona transformations.

* *Trans. American Math. Soc.* Vol. II. (1901), pp. 137–138; *Quart. Journ.* Vol. XXXIII. (1901), pp. 145–173; *Bull. American Math. Soc.* Vol. VIII. (1901), p. 63 *et sq.*; *Linear Groups*, Ch. XIV. pp. 303–307, etc.

† *Ueber die Galois'che Gruppe der Gleichung 27. Grades, von welcher die Geraden auf der allgemeinen Fläche dritter Ordnung abhängen*, Diss. Marburg. 1888.

‡ "Ueber die Galois'che Gruppe der Gleichung 28. Grades, von welcher die Doppeltangenten einer Curve vierter Ordnung abhängen," *Math. Ann.* Bd. XXIII. pp. 489–503.

§ "Groups of Substitutions connected with the Twenty-Seven Lines upon the Cubic Surface," *Annali di Matematica pura ed applicata*, Vol. XX. (1892–3), pp. 269 *et seq.*; Vol. XXI. (1893), pp. 85 *et seq.*

|| *American Journ. of Math.* Vol. XXV. No. 2 (1903), pp. 107–122.

¶ "The Cross-Ratio Group of $n!$ Cremona Transformations of Order $n-3$ in Flat Space of $n-3$ Dimensions," *American Journ. of Math.* Vol. XXII. (1900), p. 279.

** "The Cross-Ratio Group of 120 Quadratic Cremona Transformations of the Plane," *American Journ. of Math.* Vol. XXII. (1900), pp. 343–380.

INTRODUCTION

THE problem of the twenty-seven lines upon the cubic surface is of such scope and extent, and is allied to so many other problems of importance, that to give a *résumé* of all that has been done upon the subject would enlarge the present memoir into an extensive book. It has not proved feasible to attempt to cover even the geometrical phases of the problem, in their extension in particular to the cognate problem of the forty-five triple tangent planes, although the two subjects go hand in hand. In this memoir, however, is given a general survey of the problem of the twenty-seven lines, from the geometric standpoint, with special attention to salient features: the concept of trihedral pairs, the configuration of the double six, the solution of the problem of constructing models of the double six configuration and of the configurations of the straight lines upon the twenty-one types of the cubic surface, the derivation of the Pascalian configuration from that of the lines upon the cubic surface with one conical point, and certain allied problems. Certain of the results have been published, or presented before the American Mathematical Society, the North Carolina Academy of Science, and the Elisha Mitchell Scientific Society.

In §§ 1–4 are given certain preliminary theorems concerning the existence and number of the twenty-seven lines and forty-five planes for the general cubic surface, and remarks upon the first notation employed. In §§ 5–7 are given an account of Schläfli's notation, a history of the double six theorem and an analytic proof; in § 8 follow certain interesting results on the anharmonic ratios of the configurations. In § 9 appear two conditions that five lines lie upon the cubic surface, and in § 10 is the description of the formation, and the tabulation, of the thirty-six double sixes. In § 11 occur certain auxiliary theorems for special features of the general configuration of the twenty-seven lines.

In § 12 are given the definition and number of trihedral pairs, and in § 13 the actual formation of the tables of the 120 forms. In § 14 these are grouped together in such a way (sets of three) as to determine in forty ways all the twenty-seven lines. In § 15 is found a formulation of conclusions.

In § 16 is given the discussion of a particular form of the general equation of the cubic surface, together with the determination and tabulation of the forty-five triple tangent planes. In § 17 is found the formulation of the analytic expression of geometrical results.

In §§ 18–19 the methods for the construction of a model of a double six are discussed; and a practical method is there given in detail.

In §§ 20–44 the general problem of constructing thread or wire models of the configurations of the straight lines upon all twenty-one types of the cubic surface is fully considered. The complete data for constructing all these models is furnished. Drawings of the configurations of the lines, displaying their collocation with respect to each other and to the fundamental tetrahedron, have been made to scale; and these serve to illustrate the text.

In § 45 is given a discussion of the derivation of the Brianchon configuration from two spatial point triads; and in §§ 46–47 the discussion of the derivation of the Pascalian configuration from that of the straight lines upon the second species of the cubic surface (Cayley's enumeration), with a graphic representation of the combined configuration.

Finally, in § 48, appears a theorem on the number of cubic surfaces with one conical point passing through the lines of mutual intersection of two triheders.

There is appended a bibliography of the principal papers consulted which bear directly upon the present investigation.

CHAPTER I

PRELIMINARY THEOREMS

1. Existence of Straight Lines upon the Cubic Surface.

In order to find the conditions that any straight line, whose equations are

$$\frac{x - x_0}{\lambda} = \frac{y - y_0}{\mu} = \frac{z - z_0}{\nu} = r,$$

lie entirely upon a surface, we substitute

$$x = x_0 + \lambda r, \quad y = y_0 + \mu r, \quad z = z_0 + \nu r$$

in the equation of the surface; arrange the terms of the resulting equation according to powers of r and then set all the coefficients of r equal to zero, since the equation in r must be identically satisfied, i.e. for all values of r. Since in the present case the equation of the surface is of the third degree, there result four conditions. But the equations of a straight line involve four disposable constants; and, as the number of conditions to be fulfilled is exactly equal to the number of disposable constants in the equations of the straight line, it follows that every surface of the third order must contain a finite number of straight lines, real or imaginary, lying entirely upon it.

2. Number of Straight Lines upon the Cubic Surface.

Suppose we pass a plane Π through a point P outside the surface and through a straight line l lying upon the surface. Then Π meets the surface in the line l, and a conic C besides (since the curve of intersection is a degenerate cubic), i.e. meets the surface in a section having two double points. Hence, by definition, it is a double tangent plane. These double tangent planes Π to the cubic surface are also double tangent planes to the tangent cone, vertex P. Now since to every plane Π corresponds one straight line l lying entirely on the surface, and there are twenty-seven* ($n = 3$) double tangent planes to the tangent cone, vertex P, therefore there are twenty-seven straight lines l upon the cubic surface†.

3. Triple Tangent Planes.

By properly determining the plane passed through any straight line l upon the cubic surface, the conic C (§ 2) will degenerate into a pair of straight lines. Here the plane intersects the surface in three intersecting straight lines (a degenerate curve of the third order having three double points) and the points of intersection of the lines taken in pairs are the points of contact of the plane with the surface. Now, through each of the three lines in the plane there may be drawn, besides the given plane, four other triple tangent planes. For these twelve new planes give rise to twenty-four lines upon the surface, making up, with the former three lines, twenty-seven lines upon the surface. It is clear that there can be no lines upon the surface besides the twenty-seven. For since the three lines upon the triple tangent plane are the complete intersection of this plane with the surface, every other line upon the surface meets the triple tangent plane in a point upon one of the three lines, and must therefore lie in a plane passing through one of these lines, such plane (since it meets the surface in two lines, and therefore in a third line) being obviously a triple tangent plane. Hence the whole number of lines upon the surface is twenty-seven. Every straight line on the surface is met by ten others.

If all the twenty-seven intersect in pairs, there would be 351 points of intersection. But since each line is met by ten other lines, there remain sixteen lines by which it is not met. Therefore there are $\frac{27 \times 16}{2} = 216$ pairs of lines that do not mutually intersect. Consequently there are 135 points of intersection. Since these 135 points, by threes, determine the triple tangent planes, there are forty-five triple tangent planes.

* Salmon (*Geometry of Three Dimensions*, 4th edition, § 286) gives

$$\frac{n}{2}(n-1)(n-2)(n^3-n^2+n-12)$$

as the number of double tangent planes, drawn through a point P to a surface of the nth degree.

† For other proofs, cf. for example, R. Sturm, *Flächen dritter Ordnung*, Kap. 2, § 20; Cayley, *Coll. Math. Papers*, Vol. I. No. 76, pp. 445–456.

4. Salmon's Notation for the Twenty-Seven Lines*.

Lemma. *The general equation of the cubic surface may be reduced to the canonical form* $uvw - \xi\eta\zeta = 0$, *where* $u, v, w, \xi, \eta, \zeta$ *are linear polynomes.*

The number of independent constants in the general equation of the third degree is 19 $\left[\frac{n(n^2+6n+11)}{6}, \text{ for } n=3\right]$. Since the linear polynomes $u, v, w, \xi, \eta, \zeta$ contain eighteen ratios of constants and there is one other constant factor implicitly contained in one of the products uvw, $\xi\eta\zeta$, therefore the form $uvw - \xi\eta\zeta = 0$ contains nineteen constants and is one into which the general equation of a cubic surface may be thrown.

It will appear later (§ 15) from geometrical considerations that the problem to reduce the base cubic to the form $uvw - \xi\eta\zeta = 0$ is solvable in 120 different ways.

Notation. Consider the canonical form of the surface of the third degree $ace - bdf = 0$, where a, b, c, d, e, f are linear polynomes. By inspection it is patent that this surface contains the nine lines ab, ad, af, cb, cd, cf, eb, ed, ef—where ab, for example, represents the line of intersection of the planes $a=0$, $b=0$. If we suppose $a = \mu b$ to be the equation of one of the triple tangent planes through the intersection of the planes a and b, the plane $a = \mu b$ meets the surface in the same lines in which it meets the hyperboloid $\mu ce - df = 0$, that is, the two lines in the plane are generating lines of different species, and consequently one of them meets the pair of lines cd and ef, and the other of them meets the pair of lines cf and ed. Let us now denote each of the remaining eighteen lines by the three lines which it meets, the line meeting ab, cd and ef being denoted by the symbol $ab.cd.ef$. Since μ has three values, there are three lines that meet ab, cd, ef. Applying the same reasoning to the planes through bc and ca, we employ the following symbolism for the twenty-seven lines:

$$ab,\ ad,\ \ldots\ldots\ ef;$$

$$\begin{matrix} (ab.cd.ef)_i,\ (ad.cf.eb)_i,\ (af.cb.ed)_i, \\ (ab.cf.ed)_i,\ (ad.cb.ef)_i,\ (af.cd.eb)_i. \end{matrix} \qquad (i = 1, 2, 3.)$$

Unfortunately, the information furnished by this method as to how these suffixes are to be supplied is inadequate; certain postulates have to be made as to how the intersections occur. This notation of Salmon's was the first given for the twenty-seven lines. It was soon superseded by a very superior one, to be explained in the next article.

* *Camb. and Dublin Math. Journ.* Vol. iv. (1849), pp. 252–260.

CHAPTER II

THE DOUBLE SIX CONFIGURATION. AUXILIARY THEOREMS

5. The Double Six Notation.

Let us write down, in Salmon's notation, two systems of non-intersecting lines:

$$ab,\ cd,\ ef,\ (ad\,.\,cf\,.\,eb)_1,\ (ad\,.\,cf\,.\,eb)_2,\ (ad\,.\,cf\,.\,eb)_3,$$
$$cf,\ eb,\ ad,\ (ab\,.\,cd\,.\,ef)_1,\ (ab\,.\,cd\,.\,ef)_2,\ (ab\,.\,cd\,.\,ef)_3.$$

In this scheme, it is postulated (§ 4) that each line of one system does not intersect the line of the other system which is written in the same vertical line, but does intersect the five other lines of the second.

This configuration was first actually observed by Schläfli* and was given by him the name it has since borne—a "double six." The concept of the double six lies at the very basis of the study of the lines upon a cubic surface. The notation derived therefrom is the simplest and most convenient that has yet been discovered for the twenty-seven lines and forty-five planes.

Notation. Starting with the double six, written

$$a_1,\ a_2,\ a_3,\ a_4,\ a_5,\ a_6,$$
$$b_1,\ b_2,\ b_3,\ b_4,\ b_5,\ b_6,$$

we are enabled to express the complex and diversified symmetry of the twenty-seven lines and forty-five planes in unique and simple form.

* "An attempt to determine the twenty-seven lines upon a surface of the third order, and to divide such surfaces into species in reference to the reality of the lines upon the surface," *Quart. Journ.* Vol. II. (1858), pp. 55–65, 110–120.

Returning to the double six, written in Salmon's notation, we see that the lines ab, cb, and eb lie in the same plane, and are the only three of the twenty-seven lines that lie in the plane b. In like manner cb, cd, and cf all lie in the plane c, and hence the line that lies in the plane of ab and eb is identical with the line that lies in the plane of cd and cf, viz. the line cb.

In the new notation, we shall call the third line in the plane of a_1 and b_2, which intersect, the line c_{12}; and the triangle thus formed will be designated 12. As has been shown above, the side c_{12} forms with a_2 and b_1 a triangle, designated 21. Hence we have 15 ($\equiv {}_6C_2$) lines c, each of which intersects only those four lines a, b the suffixes of which belong to the pair of numbers forming the suffix of c. For suppose c_{12} should intersect any other line, say a_3, of the eight lines a_3, a_4, a_5, a_6; b_3, b_4, b_5, b_6. Then c_{12} intersecting a_1, b_1, a_2 and b_2 already, $c_{12}a_3b_1$ and $c_{12}a_3b_2$ form two triangles; and since they have two lines in common, their planes are identical, and consequently b_1 intersects b_2, contrary to hypothesis.

Any two c's, the suffixes of which have a number in common, do not intersect. For suppose c_{12}, c_{13} intersect; they form a plane in which a_1 and b_1 lie, and therefore a_1 meets b_1, contrary to hypothesis. It may also be shown that any two c's, the suffixes of which have no number in common, do intersect. These facts may be briefly put as follows:

$$\left.\begin{array}{l} c_{ij} \text{ intersects } a_i,\ b_j;\ a_j,\ b_i \\ c_{ij} \text{ intersects } c_{kl} \\ c_{ij} \text{ does not intersect } c_{ik} \\ \qquad c_{ij} \equiv c_{ji} \\ \Delta_{ij} \text{ is not identical with } \Delta_{ji} \end{array}\right\} \begin{array}{l} (i, j, k, l = 1, 2, \ldots 6, \\ i, j, k, l \text{ all distinct.}) \end{array}$$

We see then that there are triangles of the form c_{12}, c_{34}, c_{56} which may be briefly designated 12.34.56. Hence there are thirty ($\equiv {}_6C_2$) triangles of the type 12, and fifteen of the type 12.34.56. The latter arises from the fact that, if we fix our attention upon 12, the other two sets may be written in only three ways.

6. History of the Theorem.

In 1858, Schläfli (l.c.) proved the double six theorem incidentally in connection with his investigations on the twenty-seven lines on the cubic surface. He enunciated the theorem in the following form:

Given five lines a, b, c, d, e which meet the same straight line X; then may any four of the five lines be intersected by another line.

Suppose that A, B, C, D, E are the other lines intersecting (b, c, d, e), (c, d, e, a), (d, e, a, b), (e, a, b, c), *and* (a, b, c, d) *respectively. Then A, B, C, D, E will all be met by one other straight line x.*

The double six in this case is written

$$\begin{pmatrix} a, & b, & c, & d, & e, & x \\ A, & B, & C, & D, & E, & X \end{pmatrix}.$$

Schläfli then proposes the question: "Is there, for this elementary theorem, a demonstration more simple than the one derived from the theory of cubic forms?"

Sylvester* states that the theorem admits of very simple geometrical proof; but he did not supply the proof. Salmon† has given a method for constructing a double six, by pure geometry; but it is not a proof of the theorem, independent of the cubic surface.

In 1868, Cayley‡ gave a proof of the theorem from purely static considerations, making use of theorems on six lines in involution. It has recently been remarked, by Mr G. T. Bennett, that this is erroneous§. Again in 1870, Cayley‖ verified the theorem, using this time the six co-ordinates of a line. In 1903, Kasner¶ also gave a proof using the six co-ordinates of a line. More recently (January 13, 1910), Baker has given a direct algebraic proof of the theorem independently of the cubic surface, so formulated as to show that the theorem belongs to three dimensions only**.

In 1881, Schur†† originally gave a geometrical proof of the double six theorem, basing his proof on a poristic property of the plane cubic curve. Recently (November 21, 1910), Baker‡‡ has given a geometrical proof of the double six theorem independently of the cubic surface, thus demonstrating the fundamentally projective character of the configuration.

* "Note sur les 27 droites d'une surface du 3e degré," *Comptes Rendus*, Vol. LII. (1861), pp. 977–980.

† *Geometry of Three Dimensions*, 4th edition, p. 500.

‡ "A 'Smith's Prize' Paper, Solutions," *Coll. Math. Papers*, Vol. VIII. (1868), pp. 430–431.

§ "The Double Six," *Proc. London Math. Soc.* Ser. 2, Vol. IX. (1911), p. 351.

‖ "On the Double Sixers of a Cubic Surface," *Quart. Journ.* Vol. X. (1870), pp. 58–71.

¶ *American Journ. of Math.* Vol. XXV. No. 2, pp. 107–122.

** *Proc. London Math. Soc.* Ser. 2, Vol. IX. Parts II. and III. pp. 145–199.

†† *Math. Ann.* Bd. XVIII. pp. 10, 11.

‡‡ *Proc. Roy. Soc.* A, Vol. LXXXIV. pp. 597–602.

7. Proof of the Double Six Theorem.

Representing the double six as follows :

$$\begin{pmatrix} 1 & 2 & 3 & 4 & 5 & 6 \\ 1' & 2' & 3' & 4' & 5' & 6' \end{pmatrix},$$

it is seen that these twelve lines have the thirty intersections $P_{ij'}$,

	1	2	3	4	5	6
1′		•	•	•	•	•
2′	•		•	•	•	•
3′	•	•		•	•	•
4′	•	•	•		•	•
5′	•	•	•	•		•
6′	•	•	•	•	•	

and determine thirty planes $\Pi_{ij'}$ (formed by the lines i and j').

Using quadriplanar co-ordinates, I choose for the lines 1′, 3′, 4′, 5′, 6′ the following equations :

$$\begin{aligned}
1' :&\quad \delta\delta' Cx + \delta\delta' Az - (\alpha'\gamma'\delta - \alpha\gamma\delta')\, w = 0,\ y = 0,\\
3' :&\quad \gamma\gamma' Dy - (\beta'\gamma\delta' - K\beta\gamma'\delta)\, z + \gamma\gamma' Bw = 0,\ x = 0,\\
4' :&\quad z = 0,\ w = 0,\\
5' :&\quad \delta' x - \alpha' w = 0,\ \gamma' y - \beta' z = 0,\\
6' :&\quad \delta x - \alpha w = 0,\ \gamma y - \beta z = 0,
\end{aligned}$$

where we set

$$A, B, C, D \equiv (\alpha' - K\alpha), (\beta' - K\beta), (\gamma' - K\gamma), (\delta' - K\delta)$$

respectively.

These equations have been so chosen that the five lines have a common tractor*. The condition that any five lines, 1, 2, 3, 4, 5 say, have a common tractor, where the equations of lines i and j are

$$\left.\begin{aligned} a_i x + b_i y + c_i z + d_i w = 0 \\ \alpha_i x + \beta_i y + \gamma_i z + \delta_i w = 0 \end{aligned}\right\} \quad \ldots\ldots\ldots\ldots\ldots\ldots\ldots (i)$$

and

$$\left.\begin{aligned} a_j x + b_j y + c_j z + d_j w = 0 \\ \alpha_j x + \beta_j y + \gamma_j z + \delta_j w = 0 \end{aligned}\right\} \quad \ldots\ldots\ldots\ldots\ldots\ldots\ldots (j)$$

* Cayley uses the word "tractor" to denote a line which meets any given lines in space.

respectively, and we understand by (ij) the determinant

$$\begin{vmatrix} a_i, & b_i, & c_i, & d_i \\ \alpha_i, & \beta_i, & \gamma_i, & \delta_i \\ a_j, & b_j, & c_j, & d_j \\ \alpha_j, & \beta_j, & \gamma_j, & \delta_j \end{vmatrix},$$

is as follows:

$$\Delta_5 = \begin{vmatrix} 0, & (12), & (13), & (14), & (15) \\ (21), & 0, & (23), & (24), & (25) \\ (31), & (32), & 0, & (34), & (35) \\ (41), & (42), & (43), & 0, & (45) \\ (51), & (52), & (53), & (54), & 0 \end{vmatrix} \equiv 0^*.$$

The five lines 1′, 3′, 4′, 5′, 6′ are co-tractorial, since the equations identically satisfy $\Delta_5 = 0$, as may be shown on trial. Moreover these five lines do not mutually intersect, since in forming the determinants (ij), no one of them is found to vanish. A difficulty arises in the event of the hyperboloid through any three of these five collinear lines touching a fourth, that is to say, that certain four of the lines might have a double tractor†. That such is not the case will appear in the sequel.

Determining now the common tractor, 2, of these five lines, we find it to have the equations:

$$2: \quad \left.\begin{cases} \gamma' B(\delta' x - \alpha' w) - \delta' A(\gamma' y - \beta' z) = 0 \\ \gamma B(\delta x - \alpha w) - \delta A(\gamma y - \beta z) = 0 \end{cases}\right\}.$$

Now, in general, four given lines have a pair of tractors. Since the five lines 1′, 3′, 4′, 5′, 6′ already have a single tractor 2, they have, in sets of four, five more tractors thus: the lines 1, 3, 4, 5, 6 are tractors of the sets (3′, 4′, 5′, 6′), (1′, 4′, 5′, 6′), (1′, 3′, 5′, 6′), (1′, 3′, 4′, 6′), (1′, 3′, 4′, 5′) respectively.

Let us proceed to find the equations of the five lines 1, 3, 4, 5, 6. Recalling the values of A, B, C and D above, it is obvious by inspection that the equations of lines 1 and 3, meeting the quadruples (3′, 4′, 5′, 6′) and (1′, 4′, 5′, 6′) respectively, are

$$1: \quad x = 0, \quad w = 0,$$
$$3: \quad y = 0, \quad z = 0.$$

* Sylvester, "Note sur l'involution de six lignes dans l'espace," *Comptes Rendus*, Vol. LII. (1861), pp. 815–817.

† Cayley, "On the Six Co-ordinates of a Line," *Trans. Camb. Philos. Soc.* Vol. XI. Part II. (1869), pp. 290–323.

The equations of line 4, since it meets the lines 5′ and 6′, are of the form

$$\left.\begin{aligned} \frac{x}{a'} - \lambda\frac{y}{\beta'} + \lambda\frac{z}{\gamma'} - \frac{w}{\delta'} = 0 \\ \frac{x}{a} - \mu\frac{y}{\beta} + \mu\frac{z}{\gamma} - \frac{w}{\delta} = 0 \end{aligned}\right\}.$$

The conditions that this line meet the line 1′, written in the form

$$\left(\frac{x}{a'} + \frac{z}{\gamma'} - \frac{w}{\delta'}\right)\frac{1}{a\gamma} - K\left(\frac{x}{a} + \frac{z}{\gamma} - \frac{w}{\delta}\right)\frac{1}{a'\gamma'} = 0, \;\; y = 0,$$

are given by $\lambda = 1, \;\; \mu = 1.$

Then the line 4 has the equations

$$4: \left\{\begin{aligned} \frac{x}{a} - \frac{y}{\beta} + \frac{z}{\gamma} - \frac{w}{\delta} = 0 \\ \frac{x}{a'} - \frac{y}{\beta'} + \frac{z}{\gamma'} - \frac{w}{\delta'} = 0 \end{aligned}\right\},$$

and we see by inspection that this line meets the line 3′ when we write its equations in the form

$$3': \left\{\begin{aligned} \left(-\frac{y}{\beta'} + \frac{z}{\gamma'} - \frac{w}{\delta'}\right)\frac{1}{\beta\delta} - K\left(-\frac{y}{\beta} + \frac{z}{\gamma} - \frac{w}{\delta}\right)\frac{1}{\beta'\delta'} = 0 \\ x = 0 \end{aligned}\right\}.$$

Next, line 5, since it meets the lines 4′ and 6′, has equations of the form

$$\left\{\begin{aligned} z - \lambda w = 0 \\ \left(\frac{x}{a} - \frac{w}{\delta}\right) - \mu\left(\frac{y}{\beta} - \frac{z}{\gamma}\right) = 0 \end{aligned}\right\}.$$

Meeting line 3′ (see first form), it is necessary to identify the equations

$$-\frac{\mu}{\beta}y + \frac{\mu}{\gamma}z - \frac{1}{\delta}w = 0,$$

$$-\frac{1}{\delta}\left(\frac{\delta' - K\delta}{\beta' - K\beta}\right)y + \frac{(\gamma\beta'\delta' - K\gamma'\beta\delta)}{\gamma\gamma'\delta(\beta' - K\beta)}z - \frac{w}{\delta} = 0.$$

Hence $$M = \frac{\beta}{\delta}\left(\frac{\delta' - K\delta}{\beta' - K\beta}\right) = \frac{1}{\gamma'\delta}\left(\frac{\gamma\beta'\delta' - K\gamma'\beta\delta}{\beta' - K\beta}\right),$$

giving $$\gamma' : \gamma = \beta' : \beta,$$

for which $$M = \frac{\gamma}{\delta}\left(\frac{\delta' - K\delta}{\gamma' - K\gamma}\right),$$

and therefore $$\frac{\delta}{\gamma}\left(\frac{\gamma' - K\gamma}{\delta' - K\delta}\right)\left(\frac{x}{a} - \frac{w}{\delta}\right) - \left(\frac{y}{\beta} - \frac{z}{\gamma}\right) = 0.$$

Applying similar reasoning to the equation

$$z - \lambda w = 0,$$

with respect to the lines 1′ and 3′, we finally obtain

$$\lambda = \frac{\gamma'}{\delta'}.$$

Then the equations of line 5 are

$$5: \quad \left\{\begin{matrix} \delta' z - \gamma' w = 0 \\ \beta C(\delta x - \alpha w) - \alpha D(\gamma y - \beta z) = 0 \end{matrix}\right\}.$$

Determining in similar fashion the equations of line 6, we obtain

$$6: \quad \left\{\begin{matrix} \delta z - \gamma w = 0 \\ \beta' C(\delta' x - \alpha' w) - \alpha' D(\gamma' y - \beta' z) = 0 \end{matrix}\right\}.$$

It remains to show that the five lines 1, 3, 4, 5, 6 have a common tractor (in other words, are collinear).

Writing out the various determinants (ij) and substituting in the formula for Δ_5, we obtain (after reduction)

$$\Delta_5 \equiv 0.$$

Hence these five lines have a common tractor. They do not mutually intersect, since no $(ij) \equiv 0$.

Determining now the equations of the line called 2′, which meets these five lines, we find

$$2': \quad \left\{\begin{matrix} (\alpha\beta' - \alpha'\beta)\,\delta\delta'\, Cx + (\gamma\delta' - \gamma'\delta)\,\alpha\alpha'\, Bw = 0 \\ (\alpha\beta' - \alpha'\beta)\,\gamma\gamma'\, Dy + (\gamma\delta' - \gamma'\delta)\,\beta\beta'\, Az = 0 \end{matrix}\right\}.$$

Hence we reach the following conclusion, which is Schläfli's theorem :

The five lines determined from five cö-tractorial lines by choosing the remaining tractor in each set of four of the latter lines, are themselves co-tractorial.

In the above proof, the complete set of lines was derived from the five co-tractorial lines 1′, 3′, 4′, 5′, 6′, but it is immaterial from which five of the primed or unprimed lines we start. Moreover, the relation between the sets 1′, 3′, 4′, 5′, 6′ and 1, 3, 4, 5, 6 is a reversible one—the lines of one set are the tractors of the other set by fours, and *vice versa*.

8. Anharmonic Ratios.

Let us next find the co-ordinates of the points of intersection of the lines 2′, 3′, 4′, 5′, 6′ with the line 1. Determining these in the usual way and writing down also the co-ordinates of the vertex C of the fundamental tetrahedron $ABCD$, we tabulate them as follows :

$P_{12'}$:	0	$\beta\beta'(\gamma'\delta - \gamma\delta')A$	$\gamma\gamma'(\alpha\beta' - \alpha'\beta)D$	0
$P_{13'}$:	0	$\beta'\delta'\gamma - K\beta\delta\gamma'$	$\gamma\gamma' D$	0
$P_{14'}$:	0	1	0	0
$P_{15'}$:	0	β'	γ'	0
$P_{16'}$:	0	β	γ	0
C:	0	0	1	0

The anharmonic ratio of the four collinear points $P_{12'}$, $P_{13'}$, $P_{15'}$, $P_{16'}$ is identical with the anharmonic ratio of the four parameters

$$\frac{\gamma\gamma'(\alpha\beta'-\alpha'\beta)D}{\beta\beta'(\gamma'\delta-\gamma\delta')A},\ \frac{\gamma\gamma' D}{\beta'\delta'\gamma-K\beta\delta\gamma'},\ \frac{\gamma'}{\beta'},\ \frac{\gamma}{\beta}.$$

Calculating the value $\frac{\lambda_3-\lambda_1}{\lambda_3-\lambda_2}\div\frac{\lambda_4-\lambda_1}{\lambda_4-\lambda_2}$ of the anharmonic ratio of these four parameters numbered in the order in which they are written, we find

$$(P_{12'}, P_{13'}, P_{15'}, P_{16'})=\frac{\beta'\delta'}{K\beta\delta}\left\{\frac{\beta(\gamma'\delta-\gamma\delta')A-\gamma(\alpha\beta'-\alpha'\beta)D}{\beta'(\gamma'\delta-\gamma\delta')A-\gamma'(\alpha\beta'-\alpha'\beta)D}\right\}.$$

Next, let us determine the co-ordinates of the points of intersection of the lines 1, 2, 3, 5, 6 with the line 4′. These follow in the table below:

$P_{14'}$:	0	1	0	0
$P_{24'}$:	A	B	0	0
$P_{34'}$:	1	0	0	0
$P_{54'}$:	$\alpha\gamma D$	$\beta\delta C$	0	0
$P_{64'}$:	$\alpha'\gamma' D$	$\beta'\delta' C$	0	0

The anharmonic ratio of the four collinear points $P_{24'}$, $P_{34'}$, $P_{54'}$, $P_{64'}$ is identical with the anharmonic ratio of the four parameters

$$\frac{B}{A},\ 0,\ \frac{\beta\delta C}{\alpha\gamma D},\ \frac{\beta'\delta' C}{\alpha'\gamma' D}.$$

Calculating the value $\frac{\mu_3-\mu_1}{\mu_3-\mu_2}\div\frac{\mu_4-\mu_1}{\mu_4-\mu_2}$ of the anharmonic ratio of these four parameters numbered in the order in which they are written, we find

$$(P_{24'}, P_{34'}, P_{54'}, P_{64'})=\frac{\beta'\delta'}{\beta\delta}\left\{\frac{\beta\delta AC-\alpha\gamma BD}{\beta'\delta' AC-\alpha'\gamma' BD}\right\}.$$

Recalling the fact that

$$A,\ B,\ C,\ D\equiv(\alpha'-K\alpha),\ (\beta'-K\beta),\ (\gamma'-K\gamma),\ (\delta'-K\delta)$$

respectively, it is easily verified that

$$\frac{\beta'\delta'}{K\beta\delta}\left\{\frac{\beta(\gamma'\delta-\gamma\delta')A-\gamma(\alpha\beta'-\alpha'\beta)D}{\beta'(\gamma'\delta-\gamma\delta')A-\gamma'(\alpha\beta'-\alpha'\beta)D}\right\}\equiv\frac{\beta'\delta'}{\beta\delta}\left\{\frac{\beta\delta AC-\alpha\gamma BD}{\beta'\delta' AC-\alpha'\gamma' BD}\right\}.$$

Accordingly

$$\{P_{12'}, P_{13'}, P_{15'}, P_{16'}\}=\{P_{24'}, P_{34'}, P_{54'}, P_{64'}\},$$

or in briefer notation

$$(2', 3', 5', 6')_1=(2, 3, 5, 6)_{4'}.$$

Since the configuration is a symmetrical one, we have the general conclusion

$$(i_3', i_4', i_5', i_6')_{i_1}=(i_3, i_4, i_5, i_6)_{i_2'}.$$

This theorem may be stated as follows:

The anharmonic ratio of the points in which any four out of five co-tractorial lines cut the common tractor of all five is equal to the anharmonic ratio of the points where the fifth line is intersected by the correspondents of the first four.

Let us designate the anharmonic ratio of the four planes formed by the line i_1' with the lines i_3, i_4, i_5, i_6 by the symbol $(\overline{i_3\ i_4\ i_5\ i_6})_{i_1'}$. Recalling next the known theorem concerning the two tractors of four lines, viz. that the four points of either tractor and the four planes of the other tractor have the same anharmonic ratio, we obtain

$$(i_3,\ i_4,\ i_5,\ i_6)_{i_2'} = (\overline{i_3\ i_4\ i_5\ i_6})_{i_1'}.$$

Making use of the last theorem, we obtain

$$(i_3',\ i_4',\ i_5',\ i_6')_{i_1} = (\overline{i_3\ i_4\ i_5\ i_6})_{i_1'}.$$

Hence we draw the conclusion:

The anharmonic ratio of the four points, on one of five co-tractorial lines, each of which is collinear with any three of the four remaining lines, is equal to the anharmonic ratio of the four planes determined by these remaining lines and their common tractor.

9. Five Co-tractorial Lines as Primitive.

Given any five co-tractorial lines, these determine uniquely, as was shown in § 7, the double six configuration. Then if we consider the plane of ij', it will be met by the lines i', j in points which lie on the line (ij). Since ${}_6C_2 = 15$, the twelve lines of the double six together with the fifteen new lines make up twenty-seven in all, the total number upon the cubic surface*. Hence the condition $\Delta_5 = 0$ (§ 7), which is the condition that five lines be co-tractorial, is likewise the condition that five given lines lie on a cubic surface. The subject was first studied by Sylvester in connection with a theorem in the *Lehrbuch der Statik*, of Möbius (Leipzig).

If we are given five lines, defined by their six co-ordinates $(a_1,\ b_1,\ c_1,\ f_1,\ g_1,\ h_1)\dots(a_5,\ b_5,\ c_5,\ f_5,\ g_5,\ h_5)$, then the condition that these lines be co-tractorial is expressed by the equation

$$\begin{vmatrix} 0, & 12, & 13, & 14, & 15 \\ 21, & 0, & 23, & 24, & 25 \\ 31, & 32, & 0, & 34, & 35 \\ 41, & 42, & 43, & 0, & 45 \\ 51, & 52, & 53, & 54, & 0 \end{vmatrix} = 0,$$

* Sylvester, *Comptes Rendus*, Vol. LII. (1861), pp. 977–980. Cf. also Salmon, *Geometry of Three Dimensions*, 4th edition, pp. 500–501; and R. Sturm, *Flächen dritter Ordnung*, pp. 57–59.

where we set

$$a_1 f_2 + a_2 f_1 + b_1 g_2 + b_2 g_1 + c_1 h_2 + c_2 h_1 = 12, \text{ etc.*}$$

This is also the condition that these lines may lie in a cubic surface†.

The agreement between this equation of condition and that of Sylvester ($\Delta_5 = 0$ of § 7) inheres in the fact that Cayley's determinant of the fifth order above written is the square root of Sylvester's Δ_5‡.

10. Enumeration of the Double Sixes of a Cubic Surface.

It has been shown (§ 3) that the configuration of twenty-seven lines contains two hundred and sixteen pairs of non-intersecting straight lines. Each pair determines a double six; but each double six contains six pairs of such mated lines. Hence the number of double sixes is $\frac{216}{6} = 36$.

Let us next proceed to form a table of the double sixes. The original double six

$$\begin{matrix} 1, & 2, & 3, & 4, & 5, & 6 \\ 1', & 2', & 3', & 4', & 5', & 6' \end{matrix}$$

is the primitive. There is but one of this type.

Consider next the type

$$\begin{matrix} 1, & 1', & 23, & 24, & 25, & 26, \\ 2, & 2', & 13, & 14, & 15, & 16. \end{matrix}$$

The number of this type is clearly 15 ($= {}_6C_2$).

Of the type given by

$$\begin{matrix} 1, & 2, & 3, & 56, & 46, & 45, \\ 23, & 13, & 12, & 4', & 5', & 6', \end{matrix}$$

there are 20 ($= {}_6C_3$), since keeping 1, 2, 3 fixed, the remaining numerals are uniquely determined§. This concludes the enumeration, since

$$1 + 15 + 20 = 36.$$

Below is the table of all the double sixes.

* Cayley, *Coll. Math. Papers*, Vol. VII. (1867), pp. 66–98.

† Cayley, *Coll. Math. Papers*, Vol. VII. (1870), p. 178.

‡ Sylvester, *Comptes Rendus*, Vol. LII. (1861), p. 816.

§ In his paper "A Memoir on Cubic Surfaces," *Philos. Trans. Royal Soc.* Vol. CLIX. (1869), pp. 231–326, Cayley erroneously states that there are twenty of the type

$$\begin{matrix} 1, & 2, & 3, & 56, & 46, & 45, \\ 23, & 13, & 12, & 4, & 5, & 6. \end{matrix}$$

1,	2,	3,	4,	5,	6	1,	3,	6,	45,	25,	24
1′,	2′,	3′,	4′,	5′,	6′	36,	16,	13,	2′,	4′,	5′
1,	1′,	23,	24,	25,	26	1,	4,	5,	36,	26,	23
2,	2′,	13,	14,	15,	16	45,	15,	14,	2′,	3′,	6′
1,	1′,	32,	34,	35,	36	1,	4,	6,	35,	25,	23
3,	3′,	12,	14,	15,	16	46,	16,	14,	2′,	3′,	5′
1,	1′,	42,	43,	45,	46	1,	5,	6,	34,	24,	23
4,	4′,	12,	13,	15,	16	56,	16,	15,	2′,	3′,	4′
1,	1′,	52,	53,	54,	56	2,	3,	4,	56,	16,	15
5,	5′,	12,	13,	14,	16	34,	24,	23,	1′,	5′,	6′
1,	1′,	62,	63,	64,	65	2,	3,	5,	46,	16,	14
6,	6′,	12,	13,	14,	15	35,	25,	23,	1′,	4′,	6′
2,	2′,	31,	34,	35,	36	1,	2,	3,	56,	46,	45
3,	3′,	21,	24,	25,	26	23,	13,	12,	4′,	5′,	6′
2,	2′,	41,	43,	45,	46	1,	2,	4,	56,	36,	35
4,	4′,	21,	23,	25,	26	24,	14,	12,	3′,	5′,	6′
2,	2′,	51,	53,	54,	56	1,	2,	5,	46,	36,	34
5,	5′,	21,	23,	24,	26	25,	15,	12,	3′,	4′,	6′
2,	2′,	61,	63,	64,	65	1,	2,	6,	45,	35,	34
6,	6′,	21,	23,	24,	25	26,	16,	12,	3′,	4′,	5′
3,	3′,	41,	42,	45,	46	2,	3,	6,	45,	15,	14
4,	4′,	31,	32,	35,	36	36,	26,	23,	1′,	4′,	5′
3,	3′,	51,	52,	54,	56	2,	4,	5,	36,	16,	13
5,	5′,	31,	32,	34,	36	45,	25,	24,	1′,	3′,	6′
3,	3′,	61,	62,	64,	65	2,	4,	6,	35,	15,	13
6,	6′,	31,	32,	34,	35	46,	26,	24,	1′,	3′,	5′
4,	4′,	51,	52,	53,	56	2,	5,	6,	34,	14,	13
5,	5′,	41,	42,	43,	46	56,	26,	25,	1′,	3′,	4′
4,	4′,	61,	62,	63,	65	3,	4,	5,	26,	16,	12
6,	6′,	41,	42,	43,	45	45,	35,	34,	1′,	2′,	6′
5,	5′,	61,	62,	63,	64	3,	4,	6,	25,	15,	12
6,	6′,	51,	52,	53,	54	46,	36,	34,	1′,	2′,	5′
1,	3,	4,	56,	26,	25	3,	5,	6,	24,	14,	12
34,	14,	13,	2′,	5′,	6′	56,	36,	35,	1′,	2′,	4′
1,	3,	5,	46,	26,	24	4,	5,	6,	23,	13,	12
35,	15,	13,	2′,	4′,	6′	56,	46,	45,	1′,	2′,	3′

It is worthy of remark that the double sixes play a part in the theory of the nodes of the cubic surface. If a surface of the third order $f(x, y, z, w) = 0$ has a *proper node* (x, y, z, w), then the six lines passing through such node and represented by the equations $D^2 f = 0$, $D^3 f = 0$ form a double six, in which each two corresponding (non-intersecting) lines of the two sextuples coincide*.

So in Cayley's enumeration†, the system of lines and planes for the second species of cubic surface is derived from that of the first species by supposing that in the double six the corresponding lines $1, 1'$; $2, 2'$; etc. severally coincide (cf. § 25).

11. Auxiliary Theorems.

In addition to the double six configuration, there are very many others having interesting properties formed from certain parts of the complete configuration of the twenty-seven lines. Indeed, as Cayley has remarked, the number of such theorems might be multiplied indefinitely. It is possible to deduce a large number of theorems directly from the notation; or even, more cumbrously, from an intersection table (cf. figure). Below are given a few simple theorems.

I. Any straight line is not cut by sixteen other straight lines.

II. Any two non-intersecting straight lines, say a_1 and b_1, are met by the same five lines c_{12}, c_{13}, c_{14}, c_{15}, c_{16}. Of the remaining twenty straight lines, there are five which meet only a_1, five which meet only b_1, and ten which meet neither a_1 nor b_1. A set of lines such as a_1, b_1 is called a "double."

III. Any three non-intersecting straight lines a_1, a_2, a_3 are met by the same three straight lines b_4, b_5, b_6. There are six straight lines which meet neither of the three lines a_1, a_2, a_3; six which meet only two of these three, and nine which meet only one of the three. A set of lines such as a_1, a_2, a_3 is called a "triple."

IV. Any four non-intersecting straight lines a_1, a_2, a_3, a_4 are met by two straight lines b_5, b_6. There are three straight lines which meet neither of the four, a_1, a_2, a_3, a_4; four which meet only three of them, six which meet only two of them, and eight which meet only one of them. A set of lines such as a_1, a_2, a_3, a_4 is called a "quadruple."

V (1). Five non-intersecting straight lines such as a_1, a_2, a_3, a_4, a_5, which belong to a double six, are met by only one straight line, b_6. There is but one line, a_6, which fails to meet all of them.

* Schläfli, *Quart. Journ.* Vol. II. (1858), p. 120.

† Cayley, *Coll. Math. Papers*, Vol. VI. p. 383.

	a_1	a_2	a_3	a_4	a_5	a_6	b_1	b_2	b_3	b_4	b_5	b_6	c_{12}
a_1								●	●	●	●	●	●
a_2							●		●	●	●	●	●
a_3							●	●		●	●	●	
a_4							●	●	●		●	●	
a_5							●	●	●	●		●	
a_6							●	●	●	●	●		
b_1		●	●	●	●	●							●
b_2	●		●	●	●	●							●
b_3	●	●		●	●	●							
b_4	●	●	●		●	●							
b_5	●	●	●	●		●							
b_6	●	●	●	●	●								
c_{12}	●	●					●	●					
c_{13}	●		●				●		●				
c_{14}	●			●			●			●			
c_{15}	●				●		●				●		
c_{16}	●					●	●					●	
c_{23}		●	●					●	●				
c_{24}		●		●				●		●			
c_{25}		●			●			●			●		
c_{26}		●				●		●				●	
c_{34}			●	●					●	●			●
c_{35}			●		●				●		●		●
c_{36}			●			●			●			●	●
c_{45}				●	●					●	●		●
c_{46}				●		●				●		●	●
c_{56}					●	●					●	●	●

Facing p. 24

Intersecti

c_{13}	c_{14}	c_{15}	c_{16}	c_{23}	c_{24}	c_{25}	c_{26}	c_{34}	c_{35}	c_{36}	c_{45}	c_{46}	c_{56}
●	●	●	●										
				●	●	●	●						
●				●				●	●	●			
	●				●			●			●	●	
		●				●			●		●		●
			●				●			●		●	●
●	●	●	●										
				●	●	●	●						
●				●				●	●	●			
	●				●			●			●	●	
		●				●			●		●		●
			●				●			●		●	●
								●	●	●	●	●	●
					●	●	●				●	●	●
				●		●	●		●	●			●
				●	●		●	●		●		●	
				●	●	●		●	●		●		
	●	●	●								●	●	●
●		●	●						●	●			●
●	●		●					●		●		●	
●	●	●						●	●		●		
		●	●			●	●						●
	●		●		●		●					●	
	●	●			●	●					●		
●			●	●			●			●			
●		●		●		●			●				
●	●			●	●			●					

on Table.

V (2). Five non-intersecting straight lines a_1, a_2, a_3, a_4, c_{56}, not belonging to a double six, are met by two straight lines b_5, b_6. There is no straight line by which no one of the five given straight lines is met.

Either set of five non-intersecting straight lines is called a "quintuple."

VI. Finally, such a set as six non-intersecting straight lines a_1, a_2, a_3, a_4, a_5, a_6 is called a "sextuple."

On the basis of the preceding, it is easy to determine immediately the number of doubles, triples, etc. in the configuration of the twenty-seven lines.

$$\text{Number of doubles} = \frac{27.16}{1.2} = 216;$$

$$\text{,, ,, triples} = \frac{27.16.10}{1.2.3} = 720;$$

$$\text{,, ,, quadruples} = \frac{27.16.10.6}{1.2.3.4} = 1080;$$

$$\text{,, ,, quintuples} = \frac{27.16.10.6.3}{1.2.3.4.5} = 648;$$

$$\text{,, ,, sextuples} = \frac{27.16.10.6.2.1}{1.2.3.4.5.6} = 72.$$

A word must be said about the quintuples, which are of two types. Every quadruple a_1, a_2, a_3, a_4 gives (1) one quintuple with two intersecting lines; and (2) two quintuples with one intersecting line each. That is, we have the three quintuples:

$a_1a_2a_3a_4a_5$ with one intersector b_6;
$a_1a_2a_3a_4a_6$,, ,, ,, b_5;
$a_1a_2a_3a_4c_{56}$,, two intersectors b_5, b_6.

Thus the quintuples fall into two groups; and there are twice as many in one group as in the other. Since the total number is 648, it follows that there are 432 of the type having only one intersector, and 216 of the type having two intersectors. This explains the derivation of the number of sextuples, since two quintuples out of every set of three belong to a double six.

A large number of theorems upon special portions of the configuration of the twenty-seven lines is given by Steiner*, Sturm†, Taylor‡, and others; and to these the reader is referred.

* *Crelle's Journ.* Vol. LIII. (1857), pp. 133–141.

† *Math. Ann.* Bd. XXIII. (1884), pp. 289–310. Also see Sturm's work *Synthetische Untersuchungen über Flächen dritter Ordnung.*

‡ *Philos. Trans. Royal Soc.* Part I. A (1894), pp. 37–69.

CHAPTER III

THE TRIHEDRAL PAIR CONFIGURATION

12. Definition and Number of Trihedral Pairs.

Let us choose from the forty-five triangles, formed by the twenty-seven lines, two: 12, 43—having a_1, b_2, c_{12}; a_4, b_3, c_{43} for sides respectively. These triangles have no side in common, and their planes cut in another straight line K, called their edge. Moreover their sides meet in pairs a_1, b_3; a_4, b_2; c_{12}, c_{43} upon the edge K, in three points δ, say.

The pairs just written are sides of three other triangles $a_1b_3c_{13}$, $a_4b_2c_{42}$, $c_{12}c_{43}c_{56}$, written in the abbreviated notation 13, 42, 12.43.56. The planes of the triangles 12, 43, 13.42.56 form a triheder T. Since the joins of corresponding vertices are concurrent, it follows that the meets of corresponding sides lie by threes upon the three axes of the planes of the given triangles. In like manner, the planes of the triangles 13, 42, 12.43.56 form a triheder T_1, upon whose axes their sides meet each other. The latter three triangles, like the former, have the nine lines a_1, b_2, c_{12}; a_4, b_3, c_{43}; c_{13}, c_{42}, c_{56} as sides; and the planes of the two triheders T and T_1 cut each other in these same lines. Two such triheders, T and T_1, are called conjugate; and taken together, we shall refer to them as a trihedral pair (Fig. 1).

Otherwise phrased, any two triangle planes 12, 43 which have no line in common, determine a third plane 13.42.56, which forms with them a triheder. These in turn determine another triheder 13, 42, 12.43.56.

In order to determine the number of trihedral pairs directly, it suffices to fix our attention upon any one triangle plane such as $a_1b_2c_{12}$. Through each of the lines a_1, b_2, and c_{12} pass four triple tangent planes, besides the plane $(a_1b_2c_{12})$ in question, Π say. Hence there are $45-13 = 32$ planes which have no line in common with Π. Hence in order to find the number of triheders we must multiply 45 by 32, divide by 2 since the plane is considered twice in the enumeration, and further divide by 3, since it takes three planes to form a triheder. Hence

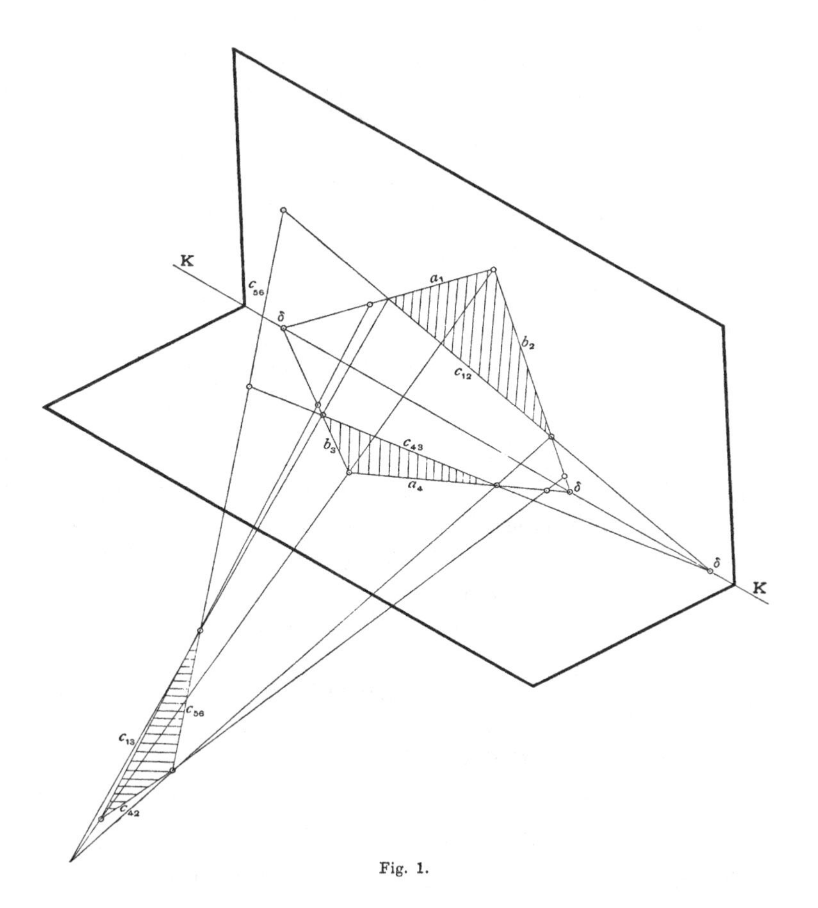

Fig. 1.

there are $\frac{45 \cdot 32}{2 \cdot 3} = 240$ triheders; but since these fall into pairs such that one triheder being given the other is uniquely determined, there are in all 120 trihedral pairs.

13. Enumeration of Trihedral Pairs.

It is of interest to form, actually, the 120 trihedral pairs in terms of the planes which constitute them. Remembering that every trihedral pair determines nine straight lines lying wholly upon the surface, we must choose for combination those triheders that fulfil this condition.

It is sufficiently obvious, in view of the unique notation already adopted, that the faces of the trihedral pair, of the type

$$(12)(23)(31) + (13)(32)(21),$$

intersect the surface in the nine lines a_1, a_2, a_3; b_1, b_2, b_3; c_{12}, c_{23}, c_{31}. These nine lines may be arranged in the following form:

$$\left\{\begin{matrix} a_1 & b_2 & c_{12} \\ a_2 & b_3 & c_{23} \\ a_3 & b_1 & c_{31} \end{matrix}\right\},$$

where each horizontal line represents three co-planar lines, and each vertical column represents three non-intersecting lines—as indicated. Bearing in mind the order of the letters, as indicated by the diagram, we may represent all examples of this type in the abbreviated notation

$$\left\{\begin{matrix} i & j & ij \\ j & k & jk \\ k & i & ki \end{matrix}\right\} i, j, k = 1, 2, \ldots 6 \ (i \neq j \neq k).$$

Fixing our attention on the first column, it is obvious that the number of this type is clearly ${}_6C_3 = 20$.

Consider next the two similar types:

$$(35)(64)(12 \cdot 34 \cdot 56) + (34)(65)(12 \cdot 35 \cdot 64);$$
$$(53)(46)(12 \cdot 34 \cdot 56) + (43)(56)(12 \cdot 35 \cdot 64).$$

The faces of the first pair cut the surface in the nine lines a_3, a_6; b_4, b_5; $c_{12}, c_{34}, c_{56}, c_{35}, c_{46}$. Also the faces of the second pair cut the surface in the nine lines a_4, a_5; b_3, b_6; $c_{12}, c_{34}, c_{56}, c_{35}, c_{46}$.

These two sets of nine lines may be arranged in the following forms:

$$\left\{\begin{matrix} a_3 & b_5 & c_{35} \\ a_6 & b_4 & c_{46} \\ c_{34} & c_{56} & c_{12} \end{matrix}\right\}, \quad \left\{\begin{matrix} a_4 & b_3 & c_{34} \\ a_5 & b_6 & c_{56} \\ c_{46} & c_{35} & c_{12} \end{matrix}\right\}.$$

That there are forty-five of each of these types follows from the fact that, if we keep 12 fixed, for example, then there are three ways in which the c-triangles may be written:

$$\left\{\begin{matrix} 35 \\ 46 \\ 34.56.12 \end{matrix}\right\}, \quad \left\{\begin{matrix} 36 \\ 45 \\ 35.46.12 \end{matrix}\right\}, \quad \left\{\begin{matrix} 34 \\ 56 \\ 36.45.12 \end{matrix}\right\}.$$

Hence there are fifteen such sets.

All examples of this type may be represented in the abbreviated notation

$$\left\{\begin{matrix} i & j & ij \\ i' & j' & i'j' \\ ij' & i'j & kl \end{matrix}\right\} i, j, k, l = 1, 2, \ldots 6 \ (i \neq j \neq k \neq l),$$

the affixes denoting that a different choice of numerals must be made for the letters in the second line to those in the first line.

Finally, there is a type:

$$(14.25.36)(35.16.24)(26.34.15)$$
$$+ (14.35.26)(25.16.34)(36.24.15).$$

The faces of this pair cut the surface in the nine lines c_{14}, c_{15}, c_{16}, c_{24}, c_{25}, c_{26}, c_{34}, c_{35}, c_{36}. These nine lines may be arranged in the following form:

$$\left\{\begin{matrix} c_{14} & c_{15} & c_{16} \\ c_{24} & c_{25} & c_{26} \\ c_{34} & c_{35} & c_{36} \end{matrix}\right\}.$$

Obviously such a form arises from the two forms:

$$\left\{\begin{matrix} 1 & 1 & 1 \\ 2 & 2 & 2 \\ 3 & 3 & 3 \end{matrix}\right\}, \quad \left\{\begin{matrix} 4 & 5 & 6 \\ 4 & 5 & 6 \\ 4 & 5 & 6 \end{matrix}\right\}.$$

Hence the number of such forms is $\frac{1}{2}({}_6C_3) = 10$.

All examples of this type may be represented in the abbreviated notation

$$\left\{\begin{matrix} ij & ik & il \\ i'j & i'k & i'l \\ i''j & i''k & i''l \end{matrix}\right\} i, j, k, l = 1, 2, \ldots 6 \ (i \neq j \neq k \neq l),$$

the affixes denoting that a different choice of numerals must be made for the letters for each line.

Hence we have enumerated all the different types, the total number of trihedral pairs being $120 = 20 + 90 + 10$.

Below are listed the 120 trihedral pairs, according to the rules just enumerated.

TABLE I

Form	Determining
(12) (23) (31) + (32) (21) (13)	$a_1, a_2, a_3; b_1, b_2, b_3; c_{12}, c_{13}, c_{23}$
(12) (24) (41) + (42) (21) (14)	$a_1, a_2, a_4; b_1, b_2, b_4; c_{12}, c_{14}, c_{24}$
(12) (25) (51) + (52) (21) (15)	$a_1, a_2, a_5; b_1, b_2, b_5; c_{12}, c_{15}, c_{25}$
(12) (26) (61) + (62) (21) (16)	$a_1, a_2, a_6; b_1, b_2, b_6; c_{12}, c_{16}, c_{26}$
(13) (34) (41) + (43) (31) (14)	$a_1, a_3, a_4; b_1, b_3, b_4; c_{13}, c_{14}, c_{34}$
(13) (35) (51) + (53) (31) (15)	$a_1, a_3, a_5; b_1, b_3, b_5; c_{13}, c_{15}, c_{35}$
(13) (36) (61) + (63) (31) (16)	$a_1, a_3, a_6; b_1, b_3, b_6; c_{13}, c_{16}, c_{36}$
(14) (45) (51) + (54) (41) (15)	$a_1, a_4, a_5; b_1, b_4, b_5; c_{14}, c_{15}, c_{45}$
(14) (46) (61) + (64) (41) (16)	$a_1, a_4, a_6; b_1, b_4, b_6; c_{14}, c_{16}, c_{46}$
(15) (56) (61) + (65) (51) (16)	$a_1, a_5, a_6; b_1, b_5, b_6; c_{15}, c_{16}, c_{56}$
(23) (34) (42) + (43) (32) (24)	$a_2, a_3, a_4; b_2, b_3, b_4; c_{23}, c_{24}, c_{34}$
(23) (35) (52) + (53) (32) (25)	$a_2, a_3, a_5; b_2, b_3, b_5; c_{23}, c_{25}, c_{35}$
(23) (36) (62) + (63) (32) (26)	$a_2, a_3, a_6; b_2, b_3, b_6; c_{23}, c_{26}, c_{36}$
(24) (45) (52) + (54) (42) (25)	$a_2, a_4, a_5; b_2, b_4, b_5; c_{24}, c_{25}, c_{45}$
(24) (46) (62) + (64) (42) (26)	$a_2, a_4, a_6; b_2, b_4, b_6; c_{24}, c_{26}, c_{46}$
(25) (56) (62) + (65) (52) (26)	$a_2, a_5, a_6; b_2, b_5, b_6; c_{25}, c_{26}, c_{56}$
(34) (45) (53) + (54) (43) (35)	$a_3, a_4, a_5; b_3, b_4, b_5; c_{34}, c_{35}, c_{45}$
(34) (46) (63) + (64) (43) (36)	$a_3, a_4, a_6; b_3, b_4, b_6; c_{34}, c_{36}, c_{46}$
(35) (56) (63) + (65) (53) (36)	$a_3, a_5, a_6; b_3, b_5, b_6; c_{35}, c_{36}, c_{56}$
(45) (56) (64) + (65) (54) (46)	$a_4, a_5, a_6; b_4, b_5, b_6; c_{45}, c_{46}, c_{56}$

TABLE II

Form	Determining
(35) (64) (12.34.56) + (34) (65) (12.35.64)	$a_3,\ a_6;\ b_4,\ b_5;\ c_{12},\ c_{34},\ c_{56},\ c_{35},\ c_{46}$
(53) (46) (12.34.56) + (43) (56) (12.35.64)	$a_4,\ a_5;\ b_3,\ b_6;\ c_{12},\ c_{34},\ c_{56},\ c_{35},\ c_{46}$
(36) (45) (12.35.64) + (35) (46) (12.36.45)	$a_3,\ a_4;\ b_5,\ b_6;\ c_{12},\ c_{35},\ c_{46},\ c_{36},\ c_{45}$
(63) (54) (12.35.64) + (53) (64) (12.36.45)	$a_5,\ a_6;\ b_3,\ b_4;\ c_{12},\ c_{35},\ c_{46},\ c_{36},\ c_{45}$
(34) (56) (12.36.45) + (36) (54) (12.34.56)	$a_3,\ a_5;\ b_4,\ b_6;\ c_{12},\ c_{36},\ c_{45},\ c_{34},\ c_{56}$
(43) (65) (12.36.45) + (63) (45) (12.34.56)	$a_4,\ a_6;\ b_3,\ b_5;\ c_{12},\ c_{36},\ c_{45},\ c_{34},\ c_{56}$
(25) (64) (13.24.56) + (24) (65) (13.25.64)	$a_2,\ a_6;\ b_4,\ b_5;\ c_{13},\ c_{24},\ c_{56},\ c_{25},\ c_{46}$
(52) (46) (13.24.56) + (42) (56) (13.25.64)	$a_4,\ a_5;\ b_2,\ b_6;\ c_{13},\ c_{24},\ c_{56},\ c_{25},\ c_{46}$
(26) (45) (13.25.64) + (25) (46) (13.26.45)	$a_2,\ a_4;\ b_5,\ b_6;\ c_{13},\ c_{25},\ c_{46},\ c_{26},\ c_{45}$
(62) (54) (13.25.64) + (52) (64) (13.26.45)	$a_5,\ a_6;\ b_2,\ b_4;\ c_{13},\ c_{25},\ c_{46},\ c_{26},\ c_{45}$
(24) (56) (13.26.45) + (26) (54) (13.24.56)	$a_2,\ a_5;\ b_4,\ b_6;\ c_{13},\ c_{26},\ c_{45},\ c_{24},\ c_{56}$
(42) (65) (13.26.45) + (62) (45) (13.24.56)	$a_4,\ a_6;\ b_2,\ b_5;\ c_{13},\ c_{26},\ c_{45},\ c_{24},\ c_{56}$
(25) (63) (14.23.56) + (23) (65) (14.25.36)	$a_2,\ a_6;\ b_3,\ b_5;\ c_{14},\ c_{23},\ c_{56},\ c_{25},\ c_{36}$
(52) (36) (14.23.56) + (32) (56) (14.25.36)	$a_3,\ a_5;\ b_2,\ b_6;\ c_{14},\ c_{23},\ c_{56},\ c_{25},\ c_{36}$
(26) (35) (14.25.36) + (25) (36) (14.26.35)	$a_2,\ a_3;\ b_5,\ b_6;\ c_{14},\ c_{25},\ c_{36},\ c_{26},\ c_{35}$

TABLE II (*Continued*)

Form	Determining
(62) (53) (14.25.36) +(52) (63) (14.26.35)	$a_5,\ a_6;\ b_2,\ b_3;\ c_{14},\ c_{25},\ c_{36},\ c_{26},\ c_{35}$
(23) (56) (14.26.35) +(26) (53) (14.23.56)	$a_2,\ a_5;\ b_3,\ b_6;\ c_{14},\ c_{26},\ c_{35},\ c_{23},\ c_{56}$
(32) (65) (14.26.35) +(62) (35) (14.23.56)	$a_3,\ a_6;\ b_2,\ b_5;\ c_{14},\ c_{26},\ c_{35},\ c_{23},\ c_{56}$
(24) (63) (15.23.46) +(23) (64) (15.24.36)	$a_2,\ a_6;\ b_3,\ b_4;\ c_{15},\ c_{23},\ c_{46},\ c_{24},\ c_{36}$
(42) (36) (15.23.46) +(32) (46) (15.24.36)	$a_3,\ a_4;\ b_2,\ b_6;\ c_{15},\ c_{23},\ c_{46},\ c_{24},\ c_{36}$
(26) (34) (15.24.36) +(24) (36) (15.26.34)	$a_2,\ a_3;\ b_4,\ b_6;\ c_{15},\ c_{24},\ c_{36},\ c_{26},\ c_{34}$
(62) (43) (15.24.36) +(42) (63) (15.26.34)	$a_4,\ a_6;\ b_2,\ b_3;\ c_{15},\ c_{24},\ c_{36},\ c_{26},\ c_{34}$
(23) (46) (15.26.34) +(26) (43) (15.23.46)	$a_2,\ a_4;\ b_3,\ b_6;\ c_{15},\ c_{26},\ c_{34},\ c_{23},\ c_{46}$
(32) (64) (15.26.34) +(62) (34) (15.23.46)	$a_3,\ a_6;\ b_2,\ b_4;\ c_{15},\ c_{26},\ c_{34},\ c_{23},\ c_{46}$
(24) (53) (16.23.45) +(23) (54) (16.24.35)	$a_2,\ a_5;\ b_3,\ b_4;\ c_{16},\ c_{23},\ c_{45},\ c_{24},\ c_{35}$
(42) (35) (16.23.45) +(32) (45) (16.24.35)	$a_3,\ a_4;\ b_2,\ b_5;\ c_{16},\ c_{23},\ c_{45},\ c_{24},\ c_{35}$
(25) (34) (16.24.35) +(24) (35) (16.25.34)	$a_2,\ a_3;\ b_4,\ b_5;\ c_{16},\ c_{24},\ c_{35},\ c_{25},\ c_{34}$
(52) (43) (16.24.35) +(42) (53) (16.25.34)	$a_4,\ a_5;\ b_2,\ b_3;\ c_{16},\ c_{24},\ c_{35},\ c_{25},\ c_{34}$
(23) (45) (16.25.34) +(25) (43) (16.23.45)	$a_2,\ a_4;\ b_3,\ b_5;\ c_{16},\ c_{25},\ c_{34},\ c_{23},\ c_{45}$
(32) (54) (16.25.34) +(52) (34) (16.23.45)	$a_3,\ a_5;\ b_2,\ b_4;\ c_{16},\ c_{25},\ c_{34},\ c_{23},\ c_{45}$

TABLE II (*Continued*)

Form	Determining
(46) (15) (23.45.16) + (45) (16) (23.46.15)	$a_1, a_4; b_5, b_6; c_{16}, c_{23}, c_{45}, c_{15}, c_{46}$
(64) (51) (23.45.16) + (54) (61) (23.46.15)	$a_5, a_6; b_1, b_4; c_{16}, c_{23}, c_{45}, c_{15}, c_{46}$
(45) (61) (23.41.56) + (41) (65) (23.45.16)	$a_4, a_6; b_1, b_5; c_{14}, c_{23}, c_{56}, c_{16}, c_{45}$
(54) (16) (23.41.56) + (14) (56) (23.45.16)	$a_1, a_5; b_4, b_6; c_{14}, c_{23}, c_{56}, c_{16}, c_{45}$
(56) (41) (23.46.51) + (46) (51) (23.56.14)	$a_4, a_5; b_1, b_6; c_{15}, c_{23}, c_{46}, c_{14}, c_{56}$
(65) (14) (23.46.51) + (64) (15) (23.56.14)	$a_1, a_6; b_4, b_5; c_{15}, c_{23}, c_{46}, c_{14}, c_{56}$
(36) (15) (24.35.16) + (35) (16) (24.36.15)	$a_1, a_3; b_5, b_6; c_{16}, c_{24}, c_{35}, c_{15}, c_{36}$
(63) (51) (24.35.16) + (53) (61) (24.36.15)	$a_5, a_6; b_1, b_3; c_{16}, c_{24}, c_{35}, c_{15}, c_{36}$
(31) (56) (24.36.15) + (36) (51) (24.13.56)	$a_3, a_5; b_1, b_6; c_{15}, c_{24}, c_{36}, c_{13}, c_{56}$
(13) (65) (24.36.15) + (63) (15) (24.13.56)	$a_1, a_6; b_3, b_5; c_{15}, c_{24}, c_{36}, c_{13}, c_{56}$
(53) (16) (24.13.56) + (13) (56) (24.35.16)	$a_1, a_5; b_3, b_6; c_{13}, c_{24}, c_{56}, c_{16}, c_{35}$
(35) (61) (24.13.56) + (31) (65) (24.35.16)	$a_3, a_6; b_1, b_5; c_{13}, c_{24}, c_{56}, c_{16}, c_{35}$
(14) (63) (25.13.46) + (13) (64) (25.14.36)	$a_1, a_6; b_3, b_4; c_{13}, c_{25}, c_{46}, c_{14}, c_{36}$
(41) (36) (25.13.46) + (31) (46) (25.14.36)	$a_3, a_4; b_1, b_6; c_{13}, c_{25}, c_{46}, c_{14}, c_{36}$
(16) (34) (25.14.36) + (14) (36) (25.16.34)	$a_1, a_3; b_4, b_6; c_{14}, c_{25}, c_{36}, c_{16}, c_{34}$

TABLE II (*Continued*)

Form	Determining
(61) (43) (25.14.36) +(41) (63) (25.16.34)	$a_4,\ a_6;\ b_1,\ b_3;\ c_{14},\ c_{25},\ c_{36},\ c_{16},\ c_{34}$
(13) (46) (25.16.34) +(16) (43) (25.13.46)	$a_1,\ a_4;\ b_3,\ b_6;\ c_{16},\ c_{25},\ c_{34},\ c_{13},\ c_{46}$
(31) (64) (25.16.34) +(61) (34) (25.13.46)	$a_3,\ a_6;\ b_1,\ b_4;\ c_{16},\ c_{25},\ c_{34},\ c_{13},\ c_{46}$
(14) (53) (26.13.45) +(13) (54) (26.14.35)	$a_1,\ a_5;\ b_3,\ b_4;\ c_{13},\ c_{26},\ c_{45},\ c_{14},\ c_{35}$
(41) (35) (26.13.45) +(31) (45) (26.14.35)	$a_3,\ a_4;\ b_1,\ b_5;\ c_{13},\ c_{26},\ c_{45},\ c_{14},\ c_{35}$
(15) (34) (26.14.35) +(14) (35) (26.15.34)	$a_1,\ a_3;\ b_4,\ b_5;\ c_{14},\ c_{26},\ c_{35},\ c_{15},\ c_{34}$
(51) (43) (26.14.35) +(41) (53) (26.15.34)	$a_4,\ a_5;\ b_1,\ b_3;\ c_{14},\ c_{26},\ c_{35},\ c_{15},\ c_{34}$
(13) (45) (26.15.34) +(15) (43) (26.13.45)	$a_1,\ a_4;\ b_3,\ b_5;\ c_{15},\ c_{26},\ c_{34},\ c_{13},\ c_{45}$
(31) (54) (26.15.34) +(51) (34) (26.13.45)	$a_3,\ a_5;\ b_1,\ b_4;\ c_{15},\ c_{26},\ c_{34},\ c_{13},\ c_{45}$
(15) (62) (34.12.56) +(12) (65) (34.15.26)	$a_1,\ a_6;\ b_2,\ b_5;\ c_{12},\ c_{34},\ c_{56},\ c_{15},\ c_{26}$
(51) (26) (34.12.56) +(21) (56) (34.15.26)	$a_2,\ a_5;\ b_1,\ b_6;\ c_{12},\ c_{34},\ c_{56},\ c_{15},\ c_{26}$
(16) (25) (34.15.26) +(15) (26) (34.16.25)	$a_1,\ a_2;\ b_5,\ b_6;\ c_{15},\ c_{26},\ c_{34},\ c_{16},\ c_{25}$
(61) (52) (34.15.26) +(51) (62) (34.16.25)	$a_5,\ a_6;\ b_1,\ b_2;\ c_{15},\ c_{26},\ c_{34},\ c_{16},\ c_{25}$
(12) (56) (34.16.25) +(16) (52) (34.12.56)	$a_1,\ a_5;\ b_2,\ b_6;\ c_{16},\ c_{25},\ c_{34},\ c_{12},\ c_{56}$
(21) (65) (34.16.25) +(61) (25) (34.12.56)	$a_2,\ a_6;\ b_1,\ b_5;\ c_{16},\ c_{25},\ c_{34},\ c_{12},\ c_{56}$

TABLE II (*Continued*)

Form	Determining
(16) (42) (35.12.46) +(12) (46) (35.16.24)	$a_1, a_4; b_2, b_6; c_{12}, c_{35}, c_{46}, c_{16}, c_{24}$
(61) (24) (35.12.46) +(21) (64) (35.16.24)	$a_2, a_6; b_1, b_4; c_{12}, c_{35}, c_{46}, c_{16}, c_{24}$
(12) (64) (35.14.26) +(14) (62) (35.12.46)	$a_1, a_6; b_2, b_4; c_{14}, c_{26}, c_{35}, c_{12}, c_{46}$
(21) (46) (35.14.26) +(41) (26) (35.12.46)	$a_2, a_4; b_1, b_6; c_{14}, c_{26}, c_{35}, c_{12}, c_{46}$
(14) (26) (35.16.24) +(16) (24) (35.14.26)	$a_1, a_2; b_4, b_6; c_{16}, c_{24}, c_{35}, c_{14}, c_{26}$
(41) (62) (35.16.24) +(61) (42) (35.14.26)	$a_4, a_6; b_1, b_2; c_{16}, c_{24}, c_{35}, c_{14}, c_{26}$
(14) (52) (36.12.45) +(12) (54) (36.14.25)	$a_1, a_5; b_2, b_4; c_{12}, c_{36}, c_{45}, c_{14}, c_{25}$
(41) (25) (36.12.45) +(21) (45) (36.14.25)	$a_2, a_4; b_1, b_5; c_{12}, c_{36}, c_{45}, c_{14}, c_{25}$
(15) (24) (36.14.25) +(14) (25) (36.15.24)	$a_1, a_2; b_4, b_5; c_{14}, c_{25}, c_{36}, c_{15}, c_{24}$
(51) (42) (36.14.25) +(41) (52) (36.15.24)	$a_4, a_5; b_1, b_2; c_{14}, c_{25}, c_{36}, c_{15}, c_{24}$
(12) (45) (36.15.24) +(15) (42) (36.12.45)	$a_1, a_4; b_2, b_5; c_{15}, c_{24}, c_{36}, c_{12}, c_{45}$
(21) (54) (36.15.24) +(51) (24) (36.12.45)	$a_2, a_5; b_1, b_4; c_{15}, c_{24}, c_{36}, c_{12}, c_{45}$
(13) (62) (45.12.36) +(12) (63) (45.13.26)	$a_1, a_6; b_2, b_3; c_{12}, c_{36}, c_{45}, c_{13}, c_{26}$
(31) (26) (45.12.36) +(21) (36) (45.13.26)	$a_2, a_3; b_1, b_6; c_{12}, c_{36}, c_{45}, c_{13}, c_{26}$
(16) (23) (45.13.26) +(13) (26) (45.16.23)	$a_1, a_2; b_3, b_6; c_{13}, c_{26}, c_{45}, c_{16}, c_{23}$

TABLE II (*Continued*)

Form	Determining
(61)(32)(45.13.26) +(31)(62)(45.16.23)	$a_3, a_6;\ b_1, b_2;\ c_{13}, c_{26}, c_{45}, c_{16}, c_{23}$
(12)(36)(45.16.23) +(16)(32)(45.12.36)	$a_1, a_3;\ b_2, b_6;\ c_{16}, c_{23}, c_{45}, c_{12}, c_{36}$
(21)(63)(45.16.23) +(61)(23)(45.12.36)	$a_2, a_6;\ b_1, b_3;\ c_{16}, c_{23}, c_{45}, c_{12}, c_{36}$
(13)(52)(46.12.35) +(12)(53)(46.13.25)	$a_1, a_5;\ b_2, b_3;\ c_{12}, c_{35}, c_{46}, c_{13}, c_{25}$
(31)(25)(46.12.35) +(21)(35)(46.13.25)	$a_2, a_3;\ b_1, b_5;\ c_{12}, c_{35}, c_{46}, c_{13}, c_{25}$
(15)(23)(46.13.25) +(13)(25)(46.15.23)	$a_1, a_2;\ b_3, b_5;\ c_{13}, c_{25}, c_{46}, c_{15}, c_{23}$
(51)(32)(46.13.25) +(31)(52)(46.15.23)	$a_3, a_5;\ b_1, b_2;\ c_{13}, c_{25}, c_{46}, c_{15}, c_{23}$
(12)(35)(46.15.23) +(15)(32)(46.12.35)	$a_1, a_3;\ b_2, b_5;\ c_{15}, c_{23}, c_{46}, c_{12}, c_{35}$
(21)(53)(46.15.23) +(51)(23)(46.12.35)	$a_2, a_5;\ b_1, b_3;\ c_{15}, c_{23}, c_{46}, c_{12}, c_{35}$
(13)(42)(56.12.34) +(12)(43)(56.13.24)	$a_1, a_4;\ b_2, b_3;\ c_{12}, c_{34}, c_{56}, c_{13}, c_{24}$
(31)(24)(56.12.34) +(21)(34)(56.13.24)	$a_2, a_3;\ b_1, b_4;\ c_{12}, c_{34}, c_{56}, c_{13}, c_{24}$
(14)(23)(56.13.24) +(13)(24)(56.14.23)	$a_1, a_2;\ b_3, b_4;\ c_{13}, c_{24}, c_{56}, c_{14}, c_{23}$
(41)(32)(56.13.24) +(31)(42)(56.14.23)	$a_3, a_4;\ b_1, b_2;\ c_{13}, c_{24}, c_{56}, c_{14}, c_{23}$
(12)(34)(56.14.23) +(14)(32)(56.12.34)	$a_1, a_3;\ b_2, b_4;\ c_{14}, c_{23}, c_{56}, c_{12}, c_{34}$
(21)(43)(56.14.23) +(41)(23)(56.12.34)	$a_2, a_4;\ b_1, b_3;\ c_{14}, c_{23}, c_{56}, c_{12}, c_{34}$

TABLE III

Form	Determining
(14.25.36) (35.16.24) (26.34.15) +(14.35.26) (25.16.34) (36.24.15)	$c_{14}, c_{15}, c_{16}, c_{24}, c_{25}, c_{26}, c_{34}, c_{35}, c_{36}$
(13.25.46) (45.16.23) (26.43.15) +(13.45.26) (25.16.43) (46.23.15)	$c_{13}, c_{15}, c_{16}, c_{23}, c_{25}, c_{26}, c_{43}, c_{45}, c_{46}$
(13.24.56) (54.16.23) (26.53.14) +(13.54.26) (24.16.53) (56.23.14)	$c_{13}, c_{14}, c_{16}, c_{23}, c_{24}, c_{26}, c_{53}, c_{54}, c_{56}$
(13.24.65) (64.15.23) (25.63.14) +(13.64.25) (24.15.63) (65.23.14)	$c_{13}, c_{14}, c_{15}, c_{23}, c_{24}, c_{25}, c_{63}, c_{64}, c_{65}$
(12.35.46) (45.16.32) (36.42.15) +(12.45.36) (35.16.42) (46.32.15)	$c_{12}, c_{15}, c_{16}, c_{32}, c_{35}, c_{36}, c_{42}, c_{45}, c_{46}$
(12.34.56) (54.16.32) (36.52.14) +(12.54.36) (34.16.52) (56.32.14)	$c_{12}, c_{14}, c_{16}, c_{32}, c_{34}, c_{36}, c_{52}, c_{54}, c_{56}$
(12.34.65) (64.15.32) (35.62.14) +(12.64.35) (34.15.62) (65.32.14)	$c_{12}, c_{14}, c_{15}, c_{32}, c_{34}, c_{35}, c_{62}, c_{64}, c_{65}$
(12.43.56) (53.16.42) (46.52.13) +(12.53.46) (43.16.52) (56.42.13)	$c_{12}, c_{13}, c_{16}, c_{42}, c_{43}, c_{46}, c_{52}, c_{53}, c_{56}$
(12.43.65) (63.15.42) (45.62.13) +(12.63.45) (43.15.62) (65.42.13)	$c_{12}, c_{13}, c_{15}, c_{42}, c_{43}, c_{45}, c_{62}, c_{63}, c_{65}$
(12.53.64) (63.14.52) (54.62.13) +(12.63.54) (53.14.62) (64.52.13)	$c_{12}, c_{13}, c_{14}, c_{52}, c_{53}, c_{54}, c_{62}, c_{63}, c_{64}$

14. The twenty-seven lines uniquely determined by three trihedral pairs.

It is inferable, since each trihedral pair gives nine lines, that it is always possible to place together three trihedral pairs so as to give all twenty-seven lines. By means of the above table it will be shown that such is the case.

Choosing two trihedral pairs from the first table, and one from the last table, such as

$$\begin{Bmatrix} 1 & 2 & 12 \\ 2 & 3 & 23 \\ 3 & 1 & 31 \end{Bmatrix}, \quad \begin{Bmatrix} 4 & 5 & 45 \\ 5 & 6 & 56 \\ 6 & 4 & 64 \end{Bmatrix}, \quad \begin{Bmatrix} 14 & 15 & 16 \\ 24 & 25 & 26 \\ 34 & 35 & 36 \end{Bmatrix},$$

we see that these determine all twenty-seven lines. There are ten sets of this type, exhausting the first and third tables.

The formation of the remaining triads is accomplished by means of the following consideration. Consider the trihedral pair

$$\begin{Bmatrix} 3 & 5 & 35 \\ 6 & 4 & 64 \\ 34 & 56 & 12 \end{Bmatrix}.$$

Obviously the pairs to be associated with this pair are those that determine a_4, a_5; b_1, b_2 and a_1, a_2 b_3, b_6—i.e. the trihedral pairs

$$\begin{Bmatrix} 4 & 1 & 41 \\ 5 & 2 & 52 \\ 42 & 51 & 36 \end{Bmatrix}, \quad \begin{Bmatrix} 1 & 5 & 15 \\ 2 & 6 & 26 \\ 16 & 25 & 34 \end{Bmatrix}.$$

An inspection of the table will reveal the truth of this statement.

It follows that, choosing any trihedral pair arbitrarily, then the two remaining pairs to be associated with this one are uniquely determined. The three triads so associated contain all the twenty lines; and it may easily be shown that it contains them twice.

Below are listed the forty triads of trihedral pairs in the abbreviated notation.

$$\left\{\begin{matrix}1 & 2 & 12\\ 2 & 3 & 23\\ 3 & 1 & 31\end{matrix}\right\},\quad \left\{\begin{matrix}4 & 5 & 45\\ 5 & 6 & 56\\ 6 & 4 & 64\end{matrix}\right\},\quad \left\{\begin{matrix}14 & 15 & 16\\ 24 & 25 & 26\\ 34 & 35 & 36\end{matrix}\right\}$$

$$\left\{\begin{matrix}1 & 2 & 12\\ 2 & 4 & 24\\ 4 & 1 & 41\end{matrix}\right\},\quad \left\{\begin{matrix}3 & 5 & 35\\ 5 & 6 & 56\\ 6 & 3 & 63\end{matrix}\right\},\quad \left\{\begin{matrix}13 & 15 & 16\\ 23 & 25 & 26\\ 43 & 45 & 46\end{matrix}\right\}$$

$$\left\{\begin{matrix}1 & 2 & 12\\ 2 & 5 & 25\\ 5 & 1 & 51\end{matrix}\right\},\quad \left\{\begin{matrix}3 & 4 & 34\\ 4 & 6 & 46\\ 6 & 3 & 63\end{matrix}\right\},\quad \left\{\begin{matrix}13 & 14 & 16\\ 23 & 24 & 26\\ 53 & 54 & 56\end{matrix}\right\}$$

$$\left\{\begin{matrix}1 & 2 & 12\\ 2 & 6 & 26\\ 6 & 1 & 61\end{matrix}\right\},\quad \left\{\begin{matrix}3 & 4 & 34\\ 4 & 5 & 45\\ 5 & 3 & 53\end{matrix}\right\},\quad \left\{\begin{matrix}13 & 14 & 15\\ 23 & 24 & 25\\ 63 & 64 & 65\end{matrix}\right\}$$

$$\left\{\begin{matrix}1 & 3 & 13\\ 3 & 4 & 34\\ 4 & 1 & 41\end{matrix}\right\},\quad \left\{\begin{matrix}2 & 5 & 25\\ 5 & 6 & 56\\ 6 & 2 & 62\end{matrix}\right\},\quad \left\{\begin{matrix}12 & 15 & 16\\ 32 & 35 & 36\\ 42 & 45 & 46\end{matrix}\right\}$$

$$\left\{\begin{matrix}1 & 3 & 13\\ 3 & 5 & 35\\ 5 & 1 & 51\end{matrix}\right\},\quad \left\{\begin{matrix}2 & 4 & 24\\ 4 & 6 & 46\\ 6 & 2 & 62\end{matrix}\right\},\quad \left\{\begin{matrix}12 & 14 & 16\\ 32 & 34 & 36\\ 52 & 54 & 56\end{matrix}\right\}$$

$$\left\{\begin{matrix}1 & 3 & 13\\ 3 & 6 & 36\\ 6 & 1 & 61\end{matrix}\right\},\quad \left\{\begin{matrix}2 & 4 & 24\\ 4 & 5 & 45\\ 5 & 2 & 52\end{matrix}\right\},\quad \left\{\begin{matrix}12 & 14 & 15\\ 32 & 34 & 35\\ 62 & 64 & 65\end{matrix}\right\}$$

$$\left\{\begin{matrix}1 & 4 & 14\\ 4 & 5 & 45\\ 5 & 1 & 51\end{matrix}\right\},\quad \left\{\begin{matrix}2 & 3 & 23\\ 3 & 6 & 36\\ 6 & 2 & 62\end{matrix}\right\},\quad \left\{\begin{matrix}12 & 13 & 16\\ 42 & 43 & 46\\ 52 & 53 & 56\end{matrix}\right\}$$

$$\left\{\begin{matrix}1 & 4 & 14\\ 4 & 6 & 46\\ 6 & 1 & 61\end{matrix}\right\},\quad \left\{\begin{matrix}2 & 3 & 23\\ 3 & 5 & 35\\ 5 & 2 & 52\end{matrix}\right\},\quad \left\{\begin{matrix}12 & 13 & 15\\ 42 & 43 & 45\\ 62 & 63 & 65\end{matrix}\right\}$$

$$\left\{\begin{matrix}1 & 5 & 15\\ 5 & 6 & 56\\ 6 & 1 & 61\end{matrix}\right\},\quad \left\{\begin{matrix}2 & 3 & 23\\ 3 & 4 & 34\\ 4 & 2 & 42\end{matrix}\right\},\quad \left\{\begin{matrix}12 & 13 & 14\\ 52 & 53 & 54\\ 62 & 63 & 64\end{matrix}\right\}$$

$$\left\{\begin{matrix}3 & 5 & 35\\ 6 & 4 & 64\\ 34 & 56 & 12\end{matrix}\right\},\quad \left\{\begin{matrix}4 & 1 & 41\\ 5 & 2 & 52\\ 42 & 51 & 36\end{matrix}\right\},\quad \left\{\begin{matrix}1 & 5 & 15\\ 2 & 6 & 26\\ 16 & 25 & 34\end{matrix}\right\}$$

$$\left\{\begin{matrix} 4 & 3 & 43 \\ 5 & 6 & 56 \\ 46 & 35 & 12 \end{matrix}\right\}, \quad \left\{\begin{matrix} 1 & 5 & 15 \\ 2 & 4 & 24 \\ 14 & 25 & 36 \end{matrix}\right\}, \quad \left\{\begin{matrix} 3 & 2 & 32 \\ 6 & 1 & 61 \\ 31 & 62 & 45 \end{matrix}\right\}$$

$$\left\{\begin{matrix} 3 & 6 & 36 \\ 4 & 5 & 45 \\ 35 & 46 & 12 \end{matrix}\right\}, \quad \left\{\begin{matrix} 5 & 2 & 52 \\ 6 & 1 & 61 \\ 51 & 26 & 34 \end{matrix}\right\}, \quad \left\{\begin{matrix} 1 & 4 & 14 \\ 2 & 3 & 23 \\ 13 & 24 & 56 \end{matrix}\right\}$$

$$\left\{\begin{matrix} 5 & 4 & 54 \\ 6 & 3 & 63 \\ 53 & 64 & 12 \end{matrix}\right\}, \quad \left\{\begin{matrix} 1 & 6 & 16 \\ 2 & 5 & 25 \\ 15 & 26 & 34 \end{matrix}\right\}, \quad \left\{\begin{matrix} 3 & 2 & 32 \\ 4 & 1 & 41 \\ 31 & 42 & 56 \end{matrix}\right\}$$

$$\left\{\begin{matrix} 3 & 6 & 36 \\ 5 & 4 & 54 \\ 34 & 56 & 12 \end{matrix}\right\}, \quad \left\{\begin{matrix} 4 & 2 & 42 \\ 6 & 1 & 61 \\ 41 & 62 & 35 \end{matrix}\right\}, \quad \left\{\begin{matrix} 1 & 5 & 15 \\ 2 & 3 & 23 \\ 13 & 25 & 46 \end{matrix}\right\}$$

$$\left\{\begin{matrix} 4 & 5 & 45 \\ 6 & 3 & 63 \\ 43 & 65 & 12 \end{matrix}\right\}, \quad \left\{\begin{matrix} 1 & 6 & 16 \\ 2 & 4 & 24 \\ 14 & 26 & 35 \end{matrix}\right\}, \quad \left\{\begin{matrix} 3 & 2 & 32 \\ 5 & 1 & 51 \\ 52 & 31 & 46 \end{matrix}\right\}$$

$$\left\{\begin{matrix} 2 & 5 & 25 \\ 6 & 4 & 64 \\ 24 & 65 & 13 \end{matrix}\right\}, \quad \left\{\begin{matrix} 4 & 3 & 43 \\ 5 & 1 & 51 \\ 41 & 53 & 26 \end{matrix}\right\}, \quad \left\{\begin{matrix} 1 & 6 & 16 \\ 3 & 2 & 32 \\ 12 & 36 & 45 \end{matrix}\right\}$$

$$\left\{\begin{matrix} 4 & 6 & 46 \\ 5 & 2 & 52 \\ 42 & 56 & 13 \end{matrix}\right\}, \quad \left\{\begin{matrix} 1 & 5 & 15 \\ 3 & 4 & 34 \\ 14 & 35 & 26 \end{matrix}\right\}, \quad \left\{\begin{matrix} 2 & 3 & 23 \\ 6 & 1 & 61 \\ 21 & 63 & 45 \end{matrix}\right\}$$

$$\left\{\begin{matrix} 2 & 6 & 26 \\ 4 & 5 & 45 \\ 25 & 46 & 13 \end{matrix}\right\}, \quad \left\{\begin{matrix} 5 & 3 & 53 \\ 6 & 1 & 61 \\ 51 & 63 & 24 \end{matrix}\right\}, \quad \left\{\begin{matrix} 1 & 4 & 14 \\ 3 & 2 & 32 \\ 12 & 34 & 56 \end{matrix}\right\}$$

$$\left\{\begin{matrix} 5 & 4 & 54 \\ 6 & 2 & 62 \\ 52 & 64 & 13 \end{matrix}\right\}, \quad \left\{\begin{matrix} 1 & 6 & 16 \\ 3 & 5 & 35 \\ 15 & 36 & 24 \end{matrix}\right\}, \quad \left\{\begin{matrix} 2 & 3 & 23 \\ 4 & 1 & 41 \\ 21 & 43 & 56 \end{matrix}\right\}$$

$$\left\{\begin{matrix} 2 & 6 & 26 \\ 5 & 4 & 54 \\ 24 & 56 & 13 \end{matrix}\right\}, \quad \left\{\begin{matrix} 4 & 3 & 43 \\ 6 & 1 & 61 \\ 41 & 63 & 25 \end{matrix}\right\}, \quad \left\{\begin{matrix} 1 & 5 & 15 \\ 3 & 2 & 32 \\ 12 & 35 & 46 \end{matrix}\right\}$$

$$\left\{\begin{matrix} 4 & 5 & 45 \\ 6 & 2 & 62 \\ 42 & 65 & 13 \end{matrix}\right\}, \quad \left\{\begin{matrix} 1 & 6 & 16 \\ 3 & 4 & 34 \\ 14 & 36 & 25 \end{matrix}\right\}, \quad \left\{\begin{matrix} 2 & 3 & 23 \\ 5 & 1 & 51 \\ 21 & 53 & 46 \end{matrix}\right\}$$

$$\left\{\begin{matrix} 2 & 5 & 25 \\ 6 & 3 & 63 \\ 23 & 65 & 14 \end{matrix}\right\}, \quad \left\{\begin{matrix} 3 & 4 & 34 \\ 5 & 1 & 51 \\ 31 & 54 & 26 \end{matrix}\right\}, \quad \left\{\begin{matrix} 1 & 6 & 16 \\ 4 & 2 & 42 \\ 12 & 46 & 35 \end{matrix}\right\}$$

$$\left\{\begin{matrix} 3 & 6 & 36 \\ 5 & 2 & 52 \\ 32 & 56 & 14 \end{matrix}\right\}, \quad \left\{\begin{matrix} 1 & 3 & 13 \\ 4 & 5 & 45 \\ 15 & 43 & 26 \end{matrix}\right\}, \quad \left\{\begin{matrix} 2 & 4 & 24 \\ 6 & 1 & 61 \\ 21 & 64 & 35 \end{matrix}\right\}$$

$$\left\{\begin{matrix} 2 & 6 & 26 \\ 3 & 5 & 35 \\ 25 & 36 & 14 \end{matrix}\right\}, \quad \left\{\begin{matrix} 5 & 4 & 54 \\ 6 & 1 & 61 \\ 51 & 64 & 23 \end{matrix}\right\}, \quad \left\{\begin{matrix} 1 & 3 & 13 \\ 4 & 2 & 42 \\ 12 & 43 & 56 \end{matrix}\right\}$$

$$\left\{\begin{matrix} 5 & 3 & 53 \\ 6 & 2 & 62 \\ 52 & 63 & 14 \end{matrix}\right\}, \quad \left\{\begin{matrix} 1 & 6 & 16 \\ 4 & 5 & 45 \\ 15 & 46 & 23 \end{matrix}\right\}, \quad \left\{\begin{matrix} 2 & 4 & 24 \\ 3 & 1 & 31 \\ 21 & 34 & 56 \end{matrix}\right\}$$

$$\left\{\begin{matrix} 2 & 6 & 26 \\ 5 & 3 & 53 \\ 23 & 56 & 14 \end{matrix}\right\}, \quad \left\{\begin{matrix} 3 & 4 & 34 \\ 6 & 1 & 61 \\ 31 & 64 & 25 \end{matrix}\right\}, \quad \left\{\begin{matrix} 1 & 5 & 15 \\ 4 & 2 & 42 \\ 12 & 45 & 36 \end{matrix}\right\}$$

$$\left\{\begin{matrix} 3 & 5 & 35 \\ 6 & 2 & 62 \\ 32 & 65 & 14 \end{matrix}\right\}, \quad \left\{\begin{matrix} 1 & 6 & 16 \\ 4 & 3 & 43 \\ 13 & 46 & 25 \end{matrix}\right\}, \quad \left\{\begin{matrix} 2 & 4 & 24 \\ 5 & 1 & 51 \\ 21 & 54 & 36 \end{matrix}\right\}$$

$$\left\{\begin{matrix} 2 & 4 & 24 \\ 6 & 3 & 63 \\ 23 & 64 & 15 \end{matrix}\right\}, \quad \left\{\begin{matrix} 3 & 5 & 35 \\ 4 & 1 & 41 \\ 31 & 45 & 26 \end{matrix}\right\}, \quad \left\{\begin{matrix} 1 & 6 & 16 \\ 5 & 2 & 52 \\ 12 & 56 & 34 \end{matrix}\right\}$$

$$\left\{\begin{matrix} 3 & 6 & 36 \\ 4 & 2 & 42 \\ 32 & 46 & 15 \end{matrix}\right\}, \quad \left\{\begin{matrix} 1 & 4 & 14 \\ 5 & 3 & 53 \\ 13 & 54 & 26 \end{matrix}\right\}, \quad \left\{\begin{matrix} 2 & 5 & 25 \\ 6 & 1 & 61 \\ 21 & 65 & 34 \end{matrix}\right\}$$

$$\left\{\begin{matrix} 2 & 6 & 26 \\ 3 & 4 & 34 \\ 24 & 36 & 15 \end{matrix}\right\}, \quad \left\{\begin{matrix} 4 & 5 & 45 \\ 6 & 1 & 61 \\ 41 & 65 & 23 \end{matrix}\right\}, \quad \left\{\begin{matrix} 1 & 3 & 13 \\ 5 & 2 & 52 \\ 12 & 53 & 46 \end{matrix}\right\}$$

$$\left\{\begin{matrix} 4 & 3 & 43 \\ 6 & 2 & 62 \\ 42 & 63 & 15 \end{matrix}\right\}, \quad \left\{\begin{matrix} 1 & 6 & 16 \\ 5 & 4 & 54 \\ 14 & 56 & 23 \end{matrix}\right\}, \quad \left\{\begin{matrix} 2 & 5 & 25 \\ 3 & 1 & 31 \\ 21 & 35 & 46 \end{matrix}\right\}$$

$$\left\{\begin{matrix} 2 & 6 & 26 \\ 4 & 3 & 43 \\ 23 & 46 & 15 \end{matrix}\right\}, \quad \left\{\begin{matrix} 3 & 5 & 35 \\ 6 & 1 & 61 \\ 31 & 65 & 24 \end{matrix}\right\}, \quad \left\{\begin{matrix} 1 & 4 & 14 \\ 5 & 2 & 52 \\ 12 & 54 & 36 \end{matrix}\right\}$$

$$\left\{\begin{matrix} 3 & 4 & 34 \\ 6 & 2 & 62 \\ 32 & 64 & 15 \end{matrix}\right\}, \quad \left\{\begin{matrix} 1 & 6 & 16 \\ 5 & 3 & 53 \\ 13 & 56 & 24 \end{matrix}\right\}, \quad \left\{\begin{matrix} 2 & 5 & 25 \\ 4 & 1 & 41 \\ 21 & 45 & 36 \end{matrix}\right\}$$

$$\left\{\begin{matrix} 2 & 4 & 24 \\ 5 & 3 & 53 \\ 23 & 54 & 16 \end{matrix}\right\}, \quad \left\{\begin{matrix} 3 & 6 & 36 \\ 4 & 1 & 41 \\ 31 & 46 & 25 \end{matrix}\right\}, \quad \left\{\begin{matrix} 1 & 5 & 15 \\ 6 & 2 & 62 \\ 12 & 65 & 34 \end{matrix}\right\}$$

$$\left\{\begin{matrix} 3 & 5 & 35 \\ 4 & 2 & 42 \\ 32 & 45 & 16 \end{matrix}\right\}, \quad \left\{\begin{matrix} 1 & 4 & 14 \\ 6 & 3 & 63 \\ 13 & 64 & 25 \end{matrix}\right\}, \quad \left\{\begin{matrix} 2 & 6 & 26 \\ 5 & 1 & 51 \\ 21 & 56 & 34 \end{matrix}\right\}$$

$$\left\{\begin{matrix} 2 & 5 & 25 \\ 3 & 4 & 34 \\ 24 & 35 & 16 \end{matrix}\right\}, \quad \left\{\begin{matrix} 4 & 6 & 46 \\ 5 & 1 & 51 \\ 41 & 56 & 23 \end{matrix}\right\}, \quad \left\{\begin{matrix} 1 & 3 & 13 \\ 6 & 2 & 62 \\ 12 & 63 & 45 \end{matrix}\right\}$$

$$\left\{\begin{matrix} 4 & 3 & 43 \\ 5 & 2 & 52 \\ 42 & 53 & 16 \end{matrix}\right\}, \quad \left\{\begin{matrix} 1 & 5 & 15 \\ 6 & 4 & 64 \\ 14 & 65 & 23 \end{matrix}\right\}, \quad \left\{\begin{matrix} 2 & 6 & 26 \\ 3 & 1 & 31 \\ 21 & 36 & 45 \end{matrix}\right\}$$

$$\left\{\begin{matrix} 2 & 5 & 25 \\ 4 & 3 & 43 \\ 23 & 45 & 16 \end{matrix}\right\}, \quad \left\{\begin{matrix} 3 & 6 & 36 \\ 5 & 1 & 51 \\ 31 & 56 & 24 \end{matrix}\right\}, \quad \left\{\begin{matrix} 1 & 4 & 14 \\ 6 & 2 & 62 \\ 12 & 64 & 35 \end{matrix}\right\}$$

$$\left\{\begin{matrix} 3 & 4 & 34 \\ 5 & 2 & 52 \\ 32 & 54 & 16 \end{matrix}\right\}, \quad \left\{\begin{matrix} 1 & 5 & 15 \\ 6 & 3 & 63 \\ 13 & 65 & 24 \end{matrix}\right\}, \quad \left\{\begin{matrix} 2 & 6 & 26 \\ 4 & 1 & 41 \\ 21 & 46 & 35 \end{matrix}\right\}$$

15. The Cubic Surface referred to a Pair of Triheders.

It follows from the preceding discussion, in connection with § 4, that the equation of the cubic surface, when written in the form

$$UVW - \lambda XYZ = 0,$$

may be said to be referred to a pair of triheders. Since there are precisely one hundred and twenty trihedral pairs, it follows that the equation of our fundamental cubic surface may be reduced to the form

$$UVW - \lambda XYZ = 0$$

in one hundred and twenty ways. That is to say, we have arrived at a geometrical solution of the algebraic problem: In how many ways may the general equation of the third degree $(x, y, z, w)^3 = 0$ be reduced to the form

$$UVW - \lambda XYZ = 0,$$

where U, V, W; X, Y, Z are linear polynomes in x, y, z, w?

CHAPTER IV

ANALYTICAL INVESTIGATION OF THE TWENTY-SEVEN LINES AND FORTY-FIVE TRIPLE TANGENT PLANES FOR THE GENERAL EQUATION OF THE CUBIC SURFACE

16. Choice of Special Form of the Equation of the Surface.

The general equation of a surface, of degree 3, class 12, may be written in the form

$$(x, y, z, w)^3 = 0,$$

x, y, z, w representing co-ordinates referred to a fundamental tetrahedron $ABCD$, where, as usual, we denote the plane of ABC by the equation $w = 0$, and so on*.

It has been shown (§ 15) that it is possible, and that in 120 ways, to express the equation of the cubic surface in the canonical form

$$UVW - kXYZ = 0,$$

where U, V, W; X, Y, Z represent polynomes of the first degree in x, y, z, w; and then the equation is said to be referred to a pair of triheders.

Let us choose for the equation of the cubic surface, with twenty-seven distinct straight lines lying upon it—to be discussed in subsequent articles—the following simple and symmetrical form †:

$$\left(\frac{x}{x_2} + \frac{y}{y_2} + \frac{z}{z_2} + \frac{w}{w_2}\right)\left(\frac{xz}{x_1 z_1} - \frac{yw}{y_1 w_1}\right) - k\left(\frac{x}{x_1} + \frac{y}{y_1} + \frac{z}{z_1} + \frac{w}{w_1}\right)\left(\frac{xz}{x_2 z_2} - \frac{yw}{y_2 w_2}\right) = 0.$$

* Cayley, *Philos. Trans. Royal Soc.* Vol. CLIX. (1869), pp. 231–326.

† This equation with specialized coefficients was chosen because the equations of a large number of the straight lines upon the surface can be determined by inspection. Moreover, it develops that the subsequent construction of a model of the twenty-seven lines is quite feasible for this form of the surface. The equation was suggested by an equation used by Cayley, *Coll. Math. Papers*, Vol. VII. pp. 316–330 (1870).

This equation may be thrown into the following form :

$$xz\,[y_1y_2w_1w_2\,(z_2-kz_1)\,x - w_1w_2\,(x_1y_2z_1 - kx_2y_1z_2)\,y + y_1y_2w_1w_2\,(x_2-kx_1)\,z$$
$$- w_1w_2\,(x_1z_1w_2 - kx_2z_2w_1)\,w] - yw\,[z_1z_2\,(x_1y_2w_2 - kx_2y_1w_1)\,x$$
$$- x_1x_2z_1z_2\,(w_1 - kw_2)\,y + x_1x_2\,(y_2z_1w_2 - ky_1z_2w_1)\,z$$
$$- x_1x_2z_1z_2\,(y_1 - ky_2)\,w] = 0.$$

We now recognize it as being in the canonical form, referred to a pair of triheders.

By inspection of the first form of the equation written above it is manifest that the thirteen lines, given by the following equations, lie wholly upon the surface :

$$2': \quad x=0, \quad y=0 \quad \ldots\ldots(f),$$

$$1: \quad x=0, \quad w=0 \quad \ldots\ldots(g),$$

$$3: \quad y=0, \quad z=0 \quad \ldots\ldots(h),$$

$$4': \quad z=0, \quad w=0 \quad \ldots\ldots(i),$$

$$5: \quad \frac{x}{x_1}+\frac{y}{y_1}=0, \quad \frac{z}{z_1}+\frac{w}{w_1}=0 \quad \ldots\ldots(j),$$

$$6: \quad \frac{x}{x_2}+\frac{y}{y_2}=0, \quad \frac{z}{z_2}+\frac{w}{w_2}=0 \quad \ldots\ldots(k),$$

$$5': \quad \frac{x}{x_2}+\frac{w}{w_2}=0, \quad \frac{y}{y_2}+\frac{z}{z_2}=0 \quad \ldots\ldots(l),$$

$$6': \quad \frac{x}{x_1}+\frac{w}{w_1}=0, \quad \frac{y}{y_1}+\frac{z}{z_1}=0 \quad \ldots\ldots(m),$$

$$12: \quad x=0, \quad \left(\frac{y}{y_2}+\frac{z}{z_2}+\frac{w}{w_2}\right)\frac{1}{y_1w_1} - k\left(\frac{y}{y_1}+\frac{z}{z_1}+\frac{w}{w_1}\right)\frac{1}{y_2w_2}=0 \quad \ldots(n),$$

$$23: \quad y=0, \quad \left(\frac{x}{x_2}+\frac{z}{z_2}+\frac{w}{w_2}\right)\frac{1}{x_1z_1} - k\left(\frac{x}{x_1}+\frac{z}{z_1}+\frac{w}{w_1}\right)\frac{1}{x_2z_2}=0 \quad \ldots(p),$$

$$34: \quad z=0, \quad \left(\frac{x}{x_2}+\frac{y}{y_2}+\frac{w}{w_2}\right)\frac{1}{y_1w_1} - k\left(\frac{x}{x_1}+\frac{y}{y_1}+\frac{w}{w_1}\right)\frac{1}{y_2w_2}=0 \quad \ldots(q),$$

$$41: \quad w=0, \quad \left(\frac{x}{x_2}+\frac{y}{y_2}+\frac{z}{z_2}\right)\frac{1}{x_1z_1} - k\left(\frac{x}{x_1}+\frac{y}{y_1}+\frac{z}{z_1}\right)\frac{1}{x_2z_2}=0 \quad \ldots(r),$$

$$56: \quad \frac{x}{x_1}+\frac{y}{y_1}+\frac{z}{z_1}+\frac{w}{w_1}=0, \quad \frac{x}{x_2}+\frac{y}{y_2}+\frac{z}{z_2}+\frac{w}{w_2}=0 \quad \ldots\ldots(s).$$

In order to reach the designation on the left, we must have recourse to the conception of the double six:

$$\begin{matrix} 1, & 2, & 3, & 4, & 5, & 6, \\ 1', & 2', & 3', & 4', & 5', & 6', \end{matrix}$$

in which no two lines in the same row intersect, but each line of the one row intersects all but the corresponding line of the other system. Moreover any two lines such as 1, 2′ lie in a plane denoted 12′; similarly the lines 1′, 2 lie in a plane denoted 1′2. These two planes meet in a line 12; and any three lines such as 12, 34, 56 meet in pairs, lying in a plane 12 . 34 . 56.

Now, considering the first eight lines written down, the following table showing intersections enables us to designate these eight lines as indicated by the notation shown at the left.

	f	i	l	m
g	+	+	+	+
h	+	+	+	+
j	+	+		+
k	+	+	+	

Considering next the line (n), it appears that it lies in the same plane as lines 1 and 2′ and hence must be the line 12. Similar reasoning holds for the lines (p), (q) and (r). Inspection of equations (j), (k) and (s) reveals the fact that the line (s) is none other than the line 56.

It remains to compute the equations of the remaining lines, fourteen in number, which lie upon the cubic. Let us first compute the remaining four lines of the double six, viz. 1′, 2, 3′ and 4. The following method immediately suggests itself.

The lines 3, 5, 6, and 12 are met by the line 2′, and by a second line 1′. This line 1′, as a line meeting 3, 5 and 6, will be given by equations of the form

$$x+\frac{x_1}{y_1}y=\phi\left(\frac{w_1}{z_1}z+w\right); \quad x+\frac{x_2}{y_2}y=\phi\left(\frac{w_2}{z_2}z+w\right).$$

Now, noting that these equations, setting therein $x=0$, become

$$\frac{z}{z_1}+\frac{w}{w_1}=\frac{x_1}{y_1w_1\phi}y; \quad \frac{z}{z_2}+\frac{w}{w_2}=\frac{x_2}{y_2w_2\phi}y,$$

we see that the condition of intersection with the line 12 gives

$$\phi = \frac{x_2 - kx_1}{w_2 - kw_1}.$$

Along with these equations may be written the resulting equation

$$z_1 z_2 (x_2 y_1 - x_1 y_2)\, y = \phi y_1 y_2 (z_2 w_1 - z_1 w_2)\, z.$$

Carrying out the computations, similarly, for 2, 3′, and 4, we may tabulate the equations of the four lines as follows:

$$1': \left\{ \begin{aligned} x + \frac{x_1}{y_1} y &= \phi \left(\frac{w_1}{z_1} z + w \right); \\ x + \frac{x_2}{y_2} y &= \phi \left(\frac{w_2}{z_2} z + w \right); \\ z_1 z_2 (x_2 y_1 - x_1 y_2)\, y &= \phi y_1 y_2 (z_2 w_1 - z_1 w_2)\, z; \\ \phi &= \frac{x_2 - kx_1}{w_2 - kw_1}. \end{aligned} \right.$$

$$2: \left\{ \begin{aligned} x + \frac{x_1}{w_1} w &= \phi \left(y + \frac{y_1}{z_1} z \right); \\ x + \frac{x_2}{w_2} w &= \phi \left(y + \frac{y_2}{z_2} z \right); \\ z_1 z_2 (x_2 w_1 - x_1 w_2)\, w &= \phi w_1 w_2 (y_2 z_1 - y_1 z_2)\, z; \\ \phi &= -\frac{x_2 - kx_1}{y_2 - ky_1}. \end{aligned} \right.$$

$$3': \left\{ \begin{aligned} -\phi \left(\frac{y_1}{x_1} x + y \right) &= z + \frac{z_1}{w_1} w; \\ -\phi \left(\frac{y_2}{x_2} x + y \right) &= z + \frac{z_2}{w_2} w; \\ \phi w_1 w_2 (x_2 y_1 - x_1 y_2)\, x &= x_1 x_2 (z_2 w_1 - z_1 w_2)\, w; \\ \phi &= \frac{z_2 - kz_1}{y_2 - ky_1}. \end{aligned} \right.$$

$$4: \left\{ \begin{aligned} \phi \left(\frac{w_1}{x_1} x + w \right) &= z + \frac{z_1}{y_1} y; \\ \phi \left(\frac{w_2}{x_2} x + w \right) &= z + \frac{z_2}{y_2} y; \\ \phi y_1 y_2 (x_2 w_1 - x_1 w_2)\, x &= x_1 x_2 (y_2 z_1 - y_1 z_2)\, y; \\ \phi &= -\frac{z_2 - kz_1}{w_2 - kw_1}. \end{aligned} \right.$$

On examination of the above equations of the seventeen lines it will appear that some of the forty-five triple tangent planes may be determined by inspection.

In the first place, the planes $x=0$, $y=0$, $z=0$, $w=0$ are the triple tangent planes 12′, 32′, 34′, 14′, since they contain the lines 1, 2′, 12; 2′, 3, 23; 3, 4′, 34; and 4′, 1, 41 respectively.

By inspection, we observe that $\frac{x}{x_2}+\frac{w}{w_2}=0$ is the equation of plane 15′, since in it lie both the lines 1 and 5′. In precisely similar manner we determine the equations of the planes 16′, 35′, 36′, 52′, 54′, 62′, 64′.

It appears, from an examination of the equations of lines 5 and 6′, that the plane 56′ has the form

$$\frac{x}{x_1}+\frac{y}{y_1}+\frac{z}{z_1}+\frac{w}{w_1}=0;$$

and similarly for the plane 65′.

Considering now the third type, 13′ for example, it is obvious that its equation is identical with the third equation written under 3′ above, since it vanishes identically for $x=0$, $w=0$; that is

$$13': \quad \phi w_1 w_2 (x_2 y_1 - x_1 y_2)\, x = x_1 x_2 (z_2 w_1 - z_1 w_2)\, w;$$

$$\left(\phi=\frac{z_2-kz_1}{y_2-ky_1}\right);$$

in this way we may determine also the equations of the planes 24′, 31′, and 42′.

Considering the fourth type, 25′ say, it is obvious that the lines 2 and 5′ both lie in the plane

$$\left(\frac{x}{x_2}+\frac{w}{w_2}\right)+\frac{y_2}{x_2}\left(\frac{x_2-kx_1}{y_2-ky_1}\right)\left(\frac{y}{y_2}+\frac{z}{z_2}\right)=0.$$

By analogous reasoning, we obtain similar equations for the planes 26′, 45′, 46′, 51′, 53′, 61′, 63′.

The remaining types are not discoverable by inspection, and direct calculation has to be resorted to. Take the plane 23′ for example, on which lie the lines 2, 3′, and 23. Any plane through 2 is of the form

$$\left(x+\frac{x_2}{w_2}w\right)-\phi\left(y+\frac{y_2}{z_2}z\right)+\lambda\left\{\left(x+\frac{x_1}{w_1}w\right)-\phi\left(y+\frac{y_1}{z_1}z\right)\right\}=0;$$

$$\phi=-\left(\frac{x_2-kx_1}{y_2-ky_1}\right).$$

Now a point on the line 3′ is

$$(0, \quad -y_2+ky_1, \quad z_2-kz_1, \quad 0).$$

Hence (after substitution and reduction) $\lambda = -\frac{kz_1}{z_2}$. Then we have

$$\left(x+\frac{x_2}{w_2}w\right)-\phi\left(y+\frac{y_2}{z_2}z\right)-\frac{kz_1}{z_2}\left\{\left(x+\frac{x_1}{w_1}w\right)-\phi\left(y+\frac{y_1}{z_1}z\right)\right\}=0,$$

which reduces to the form

$$w_1w_2(y_2-ky_1)(z_2-kz_1)\,x+w_1w_2(x_2-kx_1)(z_2-kz_1)\,y$$
$$+\,w_1w_2(x_2-kx_1)(y_2-ky_1)\,z+(y_2-ky_1)(x_2z_2w_1-kx_1z_1w_2)\,w=0.$$

In this fashion we determine, besides the plane 23′, the three other planes of like form 21′, 41′, and 43′.

There remain fifteen equations to be determined. Consider the plane 12 . 35 . 46, which passes through the lines 12, 35, and 46. Now the planes 35′, 3′5 intersect in the line 35. Hence any plane through 35 is of the form

$$\frac{y_1}{z_1}\left(\frac{z_2-kz_1}{y_2-ky_1}\right)\left(\frac{x}{x_1}+\frac{y}{y_1}\right)+\left(\frac{z}{z_1}+\frac{w}{w_1}\right)-\lambda\left(\frac{y}{y_2}+\frac{z}{z_2}\right)=0.$$

Since this plane passes through the line 12, if we place $x=0$ in it, we must identify

$$\left\{\frac{1}{z_1}\left(\frac{z_2-kz_1}{y_2-ky_1}\right)-\frac{\lambda}{y_2}\right\}y+\left(\frac{z_2-\lambda z_1}{z_1z_2}\right)z+\frac{w}{w_1}=0 \quad \text{.........(1)}$$

with $$\left(\frac{w_2-kw_1}{y_1y_2w_1w_2}\right)y+\left(\frac{y_2z_1w_2-ky_1z_2w_1}{y_1y_2z_1z_2w_1w_2}\right)z+\left(\frac{y_2-ky_1}{y_1y_2w_1w_2}\right)w=0 \quad \text{...(2)}.$$

Hence, multiplying equation (1) through by the factor $\left(\frac{y_2-ky_1}{y_1y_2w_2}\right)$, and comparing coefficients, we have the two equations of condition:

$$\left(\frac{y_2-ky_1}{y_1y_2w_2}\right)\left\{\frac{1}{z_1}\left(\frac{z_2-kz_1}{y_2-ky_1}\right)-\frac{\lambda}{y_2}\right\}=\left(\frac{w_2-kw_1}{y_1y_2w_1w_2}\right);$$

$$\left(\frac{y_2-ky_1}{y_1y_2w_2}\right)\left(\frac{z_2-\lambda z_1}{z_1z_2}\right)=\left(\frac{y_2z_1w_2-ky_1z_2w_1}{y_1y_2z_1z_2w_1w_2}\right).$$

From either one of these equations we derive the following value for the parameter:

$$\lambda=\frac{y_2}{z_1w_1}\left(\frac{z_2w_1-z_1w_2}{y_2-ky_1}\right).$$

Substituting this value of λ in the equation of the plane, we finally obtain, after reduction,

$$y_1z_2w_1(z_2-kz_1)\,x+x_1z_1z_2(w_2-kw_1)\,y$$
$$+\,x_1(y_2z_1w_2-ky_1z_2w_1)\,z+x_1z_1z_2(y_2-ky_1)\,w=0$$

as the equation of the plane 12 . 35 . 46.

Consider now the next type, of which there are two equations, 12 . 34 . 56 and 14 . 23 . 56.

It is sufficient to derive here the equation of the plane 12 . 34 . 56. Since it passes through the intersection of the two planes 12′ and 21′, its equation is of the form

$$\{\lambda+z_1z_2(y_2-ky_1)(w_2-kw_1)\}\,x+z_1z_2(x_2-kx_1)(w_2-kw_1)\,y$$
$$+\,(x_2-kx_1)(y_2z_1w_2-ky_1z_2w_1)\,z+z_1z_2(x_2-kx_1)(y_2-ky_1)\,w=0.$$

Moreover, the plane 12 . 34 . 56 also passes through the intersection of the two planes, 34′ and 43′, whose equations are as follows:

$$34': \qquad z=0;$$

$$43': \quad (z_2-kz_1)(x_1y_2w_2-kx_2y_1w_1)\,x+x_1x_2(z_2-kz_1)(w_2-kw_1)\,y$$
$$+\,x_1x_2(y_2-ky_1)(w_2-kw_1)\,z+x_1x_2(z_2-kz_1)(y_2-ky_1)\,w=0.$$

In order to identify this second form of the plane 12 . 34 . 56 with the form written above, it is obvious by inspection that we must multiply the equation 34′ through by

$$\frac{k(x_2-kx_1)(y_2z_1-y_1z_2)(z_2w_1-z_1w_2)}{(z_2-kz_1)},$$

and the equation 43′ through by

$$\frac{z_1z_2(x_2-kx_1)}{x_1x_2(z_2-kz_1)},$$

and add the two resulting equations. Comparing coefficients of the same variable, in the two forms for the equation 12 . 34 . 56, we derive by inspection the desired equation

$$12\,.\,34\,.\,56: \quad z_1z_2(x_1y_2w_2-kx_2y_1w_1)\,x+x_1x_2z_1z_2(w_2-kw_1)\,y$$
$$+\,x_1x_2(y_2z_1w_2-ky_1z_2w_1)\,z+x_1x_2z_1z_2(y_2-ky_1)\,w=0.$$

Similarly we may derive the four equations of the type 13 . 25 . 46; and likewise the equation of 13 . 24 . 56, which is unique.

The results of the investigation, the character of the derivation having been explained in detail, are given in the following tables.

$12' \equiv \pi_1$	$x=0$
$14' \equiv \pi_2$	$w=0$
$32' \equiv \pi_3$	$y=0$
$34' \equiv \pi_4$	$z=0$
$15' \equiv \pi_5$	$\frac{x}{x_2}+\frac{w}{w_2}=0$
$16' \equiv \pi_6$	$\frac{x}{x_1}+\frac{w}{w_1}=0$
$35' \equiv \pi_7$	$\frac{y}{y_2}+\frac{z}{z_2}=0$
$36' \equiv \pi_8$	$\frac{y}{y_1}+\frac{z}{z_1}=0$
$52' \equiv \pi_9$	$\frac{x}{x_1}+\frac{y}{y_1}=0$
$54' \equiv \pi_{10}$	$\frac{z}{z_1}+\frac{w}{w_1}=0$
$62' \equiv \pi_{11}$	$\frac{x}{x_2}+\frac{y}{y_2}=0$
$64' \equiv \pi_{12}$	$\frac{z}{z_2}+\frac{w}{w_2}=0$
$56' \equiv \pi_{13}$	$\frac{x}{x_1}+\frac{y}{y_1}+\frac{z}{z_1}+\frac{w}{w_1}=0$
$65' \equiv \pi_{14}$	$\frac{x}{x_2}+\frac{y}{y_2}+\frac{z}{z_2}+\frac{w}{w_2}=0$
$13' \equiv \pi_{15}$	$w_1 w_2 (z_2-kz_1)(x_2y_1-x_1y_2)\,x-x_1x_2(y_2-ky_1)(z_2w_1-z_1w_2)\,w=0$
$24' \equiv \pi_{16}$	$w_1 w_2 (x_2-kx_1)(y_2z_1-y_1z_2)\,z+z_1z_2(y_2-ky_1)(x_2w_1-x_1w_2)\,w=0$
$31' \equiv \pi_{17}$	$z_1 z_2 (w_2-kw_1)(x_2y_1-x_1y_2)\,y-y_1y_2(x_2-kx_1)(z_2w_1-z_1w_2)\,z=0$
$42' \equiv \pi_{18}$	$y_1 y_2 (z_2-kz_1)(x_2w_1-x_1w_2)\,x+x_1x_2(w_2-kw_1)(y_2z_1-y_1z_2)\,y=0$
$25' \equiv \pi_{19}$	$\left(\frac{x}{x_2}+\frac{w}{w_2}\right)+\frac{y_2}{x_2}\left(\frac{x_2-kx_1}{y_2-ky_1}\right)\left(\frac{y}{y_2}+\frac{z}{z_2}\right)=0$
$26' \equiv \pi_{20}$	$\left(\frac{x}{x_1}+\frac{w}{w_1}\right)+\frac{y_1}{x_1}\left(\frac{x_2-kx_1}{y_2-ky_1}\right)\left(\frac{y}{y_1}+\frac{z}{z_1}\right)=0$

$45' \equiv \pi_{21}$	$\left(\frac{x}{x_2}+\frac{w}{w_2}\right)+\frac{z_2}{w_2}\left(\frac{w_2-kw_1}{z_2-kz_1}\right)\left(\frac{y}{y_2}+\frac{z}{z_2}\right)=0$
$46' \equiv \pi_{22}$	$\left(\frac{x}{x_1}+\frac{w}{w_1}\right)+\frac{z_1}{w_1}\left(\frac{w_2-kw_1}{z_2-kz_1}\right)\left(\frac{y}{y_1}+\frac{z}{z_1}\right)=0$
$51' \equiv \pi_{23}$	$\left(\frac{x}{x_1}+\frac{y}{y_1}\right)+\frac{w_1}{x_1}\left(\frac{x_2-kx_1}{w_2-kw_1}\right)\left(\frac{z}{z_1}+\frac{w}{w_1}\right)=0$
$53' \equiv \pi_{24}$	$\left(\frac{x}{x_1}+\frac{y}{y_1}\right)+\frac{z_1}{y_1}\left(\frac{y_2-ky_1}{z_2-kz_1}\right)\left(\frac{z}{z_1}+\frac{w}{w_1}\right)=0$
$61' \equiv \pi_{25}$	$\left(\frac{x}{x_2}+\frac{y}{y_2}\right)+\frac{w_2}{x_2}\left(\frac{x_2-kx_1}{w_2-kw_1}\right)\left(\frac{z}{z_2}+\frac{w}{w_2}\right)=0$
$63' \equiv \pi_{26}$	$\left(\frac{x}{x_2}+\frac{y}{y_2}\right)+\frac{z_2}{y_2}\left(\frac{y_2-ky_1}{z_2-kz_1}\right)\left(\frac{z}{z_2}+\frac{w}{w_2}\right)=0$
$21' \equiv \pi_{27}$	$z_1z_2(y_2-ky_1)(w_2-kw_1)x+z_1z_2(x_2-kx_1)(w_2-kw_1)y$ $+(x_2-kx_1)(y_2z_1w_2-ky_1z_2w_1)z+z_1z_2(x_2-kx_1)(y_2-ky_1)w=0$
$23' \equiv \pi_{28}$	$w_1w_2(y_2-ky_1)(z_2-kz_1)x+w_1w_2(x_2-kx_1)(z_2-kz_1)y$ $+w_1w_2(x_2-kx_1)(y_2-ky_1)z+(y_2-ky_1)(x_2z_2w_1-kx_1z_1w_2)w=0$
$41' \equiv \pi_{29}$	$y_1y_2(z_2-kz_1)(w_2-kw_1)x+(w_2-kw_1)(x_2y_1z_2-kx_1y_2z_1)y$ $+y_1y_2(x_2-kx_1)(w_2-kw_1)z+y_1y_2(x_2-kx_1)(z_2-kz_1)w=0$
$43' \equiv \pi_{30}$	$(z_2-kz_1)(x_1y_2w_2-kx_2y_1w_1)x+x_1x_2(z_2-kz_1)(w_2-kw_1)y$ $+x_1x_2(y_2-ky_1)(w_2-kw_1)z+x_1x_2(z_2-kz_1)(y_2-ky_1)w=0$
12.36.45 $\equiv \pi_{31}$	$y_2z_1w_2(z_2-kz_1)x+x_2z_1z_2(w_2-kw_1)y$ $+x_2(y_2z_1w_2-ky_1z_2w_1)z+x_2z_1z_2(y_2-ky_1)w=0$
12.35.46 $\equiv \pi_{32}$	$y_1z_2w_1(z_2-kz_1)x+x_1z_1z_2(w_2-kw_1)y$ $+x_1(y_2z_1w_2-ky_1z_2w_1)z+x_1z_1z_2(y_2-ky_1)w=0$
14.25.36 $\equiv \pi_{33}$	$y_1y_2w_2(z_2-kz_1)x+w_2(x_2y_1z_2-kx_1y_2z_1)y$ $+y_1y_2w_2(x_2-kx_1)z+x_2y_1z_2(y_2-ky_1)w=0$
14.26.35 $\equiv \pi_{34}$	$y_1y_2w_1(z_2-kz_1)x+w_1(x_2y_1z_2-kx_1y_2z_1)y$ $+y_1y_2w_1(x_2-kx_1)z+x_1y_2z_1(y_2-ky_1)w=0$
15.23.46 $\equiv \pi_{35}$	$y_1w_1w_2(z_2-kz_1)x+x_1z_1w_2(w_2-kw_1)y$ $+y_1w_1w_2(x_2-kx_1)z+y_1(x_2z_2w_1-kx_1z_1w_2)w=0$
15.26.34 $\equiv \pi_{36}$	$z_1(x_1y_2w_2-kx_2y_1w_1)x+x_1x_2z_1(w_2-kw_1)y$ $+x_2y_1w_1(x_2-kx_1)z+x_1x_2z_1(y_2-ky_1)w=0$
16.23.45 $\equiv \pi_{37}$	$y_2w_1w_2(z_2-kz_1)x+x_2z_2w_1(w_2-kw_1)y$ $+y_2w_1w_2(x_2-kx_1)z+y_2(x_2z_2w_1-kx_1z_1w_2)w=0$

16.25.34 $\equiv \pi_{38}$	$z_2(x_1y_2w_2-kx_2y_1w_1)\,x+x_1x_2z_2(w_2-kw_1)\,y$ $+x_1y_2w_2(x_2-kx_1)\,z+x_1x_2z_2(y_2-ky_1)\,w=0$
12.34.56 $\equiv \pi_{39}$	$z_1z_2(x_1y_2w_2-kx_2y_1w_1)\,x+x_1x_2z_1z_2(w_2-kw_1)\,y$ $+x_1x_2(y_2z_1w_2-ky_1z_2w_1)\,z+x_1x_2z_1z_2(y_2-ky_1)\,w=0$
14.23.56 $\equiv \pi_{40}$	$y_1y_2w_1w_2(z_2-kz_1)\,x+w_1w_2(x_2y_1z_2-kx_1y_2z_1)\,y$ $+y_1y_2w_1w_2(x_2-kx_1)\,z+y_1y_2(x_2z_2w_1-kx_1z_1w_2)\,w=0$
13.25.46 $\equiv \pi_{41}$	$y_1z_2w_1w_2(z_2-kz_1)(x_1y_2-x_2y_1)\,x$ $+x_1z_1z_2w_2(w_2-kw_1)(x_1y_2-x_2y_1)\,y$ $-x_1y_1y_2w_2(x_2-kx_1)(z_1w_2-z_2w_1)\,z$ $-x_1x_2y_1z_2(y_2-ky_1)(z_1w_2-z_2w_1)\,w=0$
13.26.45 $\equiv \pi_{42}$	$y_2z_1w_1w_2(z_2-kz_1)(x_1y_2-x_2y_1)\,x$ $+x_2z_1z_2w_1(w_2-kw_1)(x_1y_2-x_2y_1)\,y$ $-x_2y_1y_2w_1(x_2-kx_1)(z_1w_2-z_2w_1)\,z$ $-x_1x_2y_2z_1(y_2-ky_1)(z_1w_2-z_2w_1)\,w=0$
15.24.36 $\equiv \pi_{43}$	$y_1y_2z_1w_2(z_2-kz_1)(x_1w_2-x_2w_1)\,x$ $+x_1x_2z_1w_2(w_2-kw_1)(y_1z_2-y_2z_1)\,y$ $+x_2y_1w_1w_2(x_2-kx_1)(y_1z_2-y_2z_1)\,z$ $+x_2y_1z_1z_2(y_2-ky_1)(x_1w_2-x_2w_1)\,w=0$
16.24.35 $\equiv \pi_{44}$	$y_1y_2z_2w_1(z_2-kz_1)(x_1w_2-x_2w_1)\,x$ $+x_1x_2z_2w_1(w_2-kw_1)(y_1z_2-y_2z_1)\,y$ $+x_1y_2w_1w_2(x_2-kx_1)(y_1z_2-y_2z_1)\,z$ $+x_1y_2z_1z_2(y_2-ky_1)(x_1w_2-x_2w_1)\,w=0$
13.24.56 $\equiv \pi_{45}$	$y_1y_2z_1z_2w_1w_2(z_2-kz_1)(x_1y_2-x_2y_1)(x_1w_2-x_2w_1)\,x$ $+x_1x_2z_1z_2w_1w_2(w_2-kw_1)(x_1y_2-x_2y_1)(y_1z_2-y_2z_1)\,y$ $-x_1x_2y_1y_2w_1w_2(x_2-kx_1)(y_1z_2-y_2z_1)(z_1w_2-z_2w_1)\,z$ $-x_1x_2y_1y_2z_1z_2(y_2-ky_1)(x_1w_2-x_2w_1)(z_1w_2-z_2w_1)\,w=0$

It is clear, from inspection of the equations of the forty-five triple tangent planes just tabulated, that a perfectly symmetrical system may be derived by setting $k=1$. But the k is retained at this time for a reason that will appear in the sequel. In fact, in the construction of the models, the circumstance that k is within our choice enables us, after assigning numerical values to the other constants, to assign such a value to k as will bring all the twenty-seven lines within easy reach. In a word, we use k, so to speak, as a lever.

17. The Analytic Expression of the Equation of a Cubic Surface in One Hundred and Twenty Forms of the type UVW + λXYZ = O.

By comparing the tables of the trihedral pairs (§ 13) with the table of triple tangent planes, we may determine at once the analytic expressions of the equation of the cubic surface into 120 different forms of the type

$$UVW + \lambda XYZ = 0,$$

where U, V, W, X, Y, Z are linear polynomes in x, y, z and w.

On reverting to the second form of the equation of the surface (§ 16), the constant λ appears to be -1. It may, however, also equal $+1$, owing to the fact that the equations of the forty-five triple tangent planes have all been written with the variables arranged in the order x, y, z, w. For example,

$$\pi_1 \cdot \pi_{28} \cdot \pi_{17} - \pi_3 \cdot \pi_{27} \cdot \pi_{15} = 0$$

and

$$\pi_1 \cdot \pi_{16} \cdot \pi_{29} + \pi_{18} \cdot \pi_{27} \cdot \pi_2 = 0$$

are both forms for the equation.

Regarding the equations of the planes as absolutely fixed in sign, then the sign to be given to λ $(= \pm 1)$ can easily be determined—and indeed by inspection. It is unnecessary to list these forms.

CHAPTER V

THE CONSTRUCTION OF A MODEL OF A DOUBLE SIX

18. Historical Introduction.

The construction of a model of a double six is a subject that has interested both Sylvester and Cayley. Sylvester* sketched a geometrical construction, not only for a double six, but also for the complete configuration of the twenty-seven lines. This same construction is also given by Salmon in his *Geometry of Three Dimensions* (4th edition, p. 500). This purely geometric construction, while it is of great theoretical interest, does not on its face afford any practical method of actually making a model of the configuration. Cayley† gave a verification of Schläfli's theorem by using his method of the six co-ordinates of a line, and thereby obtained the equational representation of the twelve lines of a double six. He chose arbitrarily the four sides $1'$, 2, $3'$, 4 of a skew quadrilateral; then selecting the co-ordinates of the lines $2'$, $5'$, $6'$ in the manner conditioned by the hypotheses for intersection, he determined the six co-ordinates of the five remaining lines $4'$, 1, 3, 5, and 6. Replacing the constants by numerical values, he obtained data for the construction of a model, but found on trial that it could not be constructed successfully with the values assigned.

Later, in 1873, Cayley‡ again treated the problem, this time approaching it through the medium of the cubic surface, the historical method of approach (cf. § 1). From a specialized form of the general equation of the cubic surface, he derived the equations of the twelve lines of the double six. Unfortunately, the numerical values he selected for the constants involved were ill chosen; and when he constructed a model of the configuration, he met with only partial success, some of the lines and intersections falling beyond convenient limits.

* *Comptes Rendus*, Vol. LII. (1861), pp. 977–980.

† "On the Double-Sixers of a Cubic Surface," *Coll. Math. Papers*, Vol. VII. (1870), pp. 316–330.

‡ "On Dr Wiener's Model of a Cubic Surface with Twenty-seven Real Lines; and on the Construction of a Double-Sixer," *Coll. Math. Papers*, Vol. VIII. pp. 366–384.

More recently, Blythe* has given a very elegant method of constructing a double six, by employing five pairs of points in involution on the line of intersection of two planes, making any convenient angle with each other.

19. A Practical Construction for the Model of a Double Six.

By making use of the equations of the lines used in the proof of Schläfli's Theorem (§ 7), it is possible to construct a perfect and simple model of a double six, which clearly brings out the salient points of the configuration.

The equations used in § 7, after some slight changes for the sake of symmetry, are as follows:

$$1: \quad x=0, \quad w=0,$$

$$2: \quad \left.\begin{cases} z_2 B\,(w_2 x + x_2 w) - w_2 A\,(z_2 y + y_2 z) = 0 \\ z_1 B\,(w_1 x + x_1 w) - w_1 A\,(z_1 y + y_1 z) = 0 \end{cases}\right\},$$

$$3: \quad y=0, \quad z=0,$$

$$4: \quad \left.\begin{cases} x/x_1 + y/y_1 + z/z_1 + w/w_1 = 0 \\ x/x_2 + y/y_2 + z/z_2 + w/w_2 = 0 \end{cases}\right\},$$

$$5: \quad \left.\begin{cases} w_2 z + z_2 w = 0 \\ y_1 C\,(w_1 x + x_1 w) - x_1 D\,(z_1 y + y_1 z) = 0 \end{cases}\right\},$$

$$6: \quad \left.\begin{cases} w_1 z + z_1 w = 0 \\ y_2 C\,(w_2 x + x_2 w) - x_2 D\,(z_2 y + y_2 z) = 0 \end{cases}\right\},$$

$$1': \quad \left.\begin{cases} y = 0 \\ w_1 w_2 C x + w_1 w_2 A z + (x_2 z_2 w_1 - x_1 z_1 w_2)\, w = 0 \end{cases}\right\},$$

$$2': \quad \left.\begin{cases} (x_1 y_2 - x_2 y_1)\, w_1 w_2 C x + (z_1 w_2 - z_2 w_1)\, x_1 x_2 B w = 0 \\ (x_1 y_2 - x_2 y_1)\, z_1 z_2 D y + (z_1 w_2 - z_2 w_1)\, y_1 y_2 A z = 0 \end{cases}\right\},$$

$$3': \quad \left.\begin{cases} x = 0 \\ z_1 z_2 D y - (y_2 z_1 w_2 - k y_1 z_2 w_1)\, z + z_1 z_2 B w = 0 \end{cases}\right\},$$

$$4': \quad z=0, \quad w=0,$$

$$5': \quad \left.\begin{cases} w_2 x + x_2 w = 0 \\ z_2 y + y_2 z = 0 \end{cases}\right\},$$

$$6': \quad \left.\begin{cases} w_1 x + x_1 w = 0 \\ z_1 y + y_1 z = 0 \end{cases}\right\},$$

where we set

$$A,\ B,\ C,\ D \equiv (x_2 - kx_1),\ -(y_2 - ky_1),\ (z_2 - kz_1),\ -(w_2 - kw_1)$$

respectively.

* "To Place a Double Six in Position," *Quart. Journ.* Vol. XXXIV. No. 1 (1902), pp. 73–74.

By inspection, we note that the three lines 1, 3 and 4′ coincide with the edges *BC*, *DA* and *AB* respectively of the fundamental tetrahedron *ABCD*. Let us choose the constants as follows:

$$x_1 = 3,\ y_1 = -4,\ z_1 = 5,\ w_1 = -6;$$
$$x_2 = z_2 = 1;\ y_2 = w_2 = -1.$$

If now we make, using suitable constants, proper drawings of the planes *ABD* and *BDC*, inspection of the figure shows, as was verified by trial, that a suitable value for the constant k is $-\frac{3}{16}$. Substituting these numerical values in the equations of the twelve lines and determining for each line (except the lines 1, 3, 3′, and 4′) the co-ordinates of the points where it meets the planes of *ABD* and *BDC* respectively, we obtain the following results in tabulated form:

		x	y	z	w	Co-ordinates, for edge = 100
1 is line	*BC*	0			0	
3 ,, ,,	*DA*		0	0		
4′ ,, ,,	*AB*			0	0	
2 meets	*AB*	25	28	0	0	$x=47{\cdot}2,\ y=52{\cdot}8$
,,	*BCD*	0	252	420	150	$y=26{\cdot}6,\ z=44{\cdot}3,\ w=15{\cdot}7$
4 ,,	*ABD*	1	2	0	−1	$x=43{\cdot}3,\ y=86{\cdot}6,\ w=-43{\cdot}3$
,,	*BCD*	0	2	5	3	$y=17{\cdot}3,\ z=43{\cdot}3,\ w=26$
5 ,,	*AB*	85	124	0	0	$x=40{\cdot}7,\ y=59{\cdot}3$
,,	*BCD*	0	6	85	85	$y=3,\ z=41{\cdot}8,\ w=41{\cdot}8$
6 ,,	*AB*	34	31	0	0	$x=52{\cdot}3,\ y=47{\cdot}7$
,,	*BCD*	0	−24	255	306	$y=-3{\cdot}8,\ z=41{\cdot}1,\ w=49{\cdot}3$
1′ ,,	*CD*	0	0	47	50	$z=48{\cdot}5,\ w=51{\cdot}5$
,,	*AD*	47	0	0	62	$x=43{\cdot}1,\ w=56{\cdot}9$
2′ ,,	*BC*	0	−10	17	0	$y=-143,\ z=243$
,,	*AD*	−14	0	0	31	$x=-82{\cdot}3,\ w=182{\cdot}3$
3′ ,,	*BC*	0	76	85	0	$y=47{\cdot}2,\ z=52{\cdot}8$
,,	*CD*	0	0	35	38	$z=48,\ w=52$
5′ ,,	*BC*	0	1	1	0	$y=50,\ z=50$
,,	*AD*	1	0	0	1	$x=50,\ w=50$
6′ ,,	*BC*	0	4	5	0	$y=44{\cdot}4,\ z=55{\cdot}6$
,,	*AD*	3	0	0	6	$x=33{\cdot}3,\ w=66{\cdot}7$

For the actual details of the construction, I have for convenience taken x, y, z, w as the perpendicular distances of the current point from the faces of a regular tetrahedron, the edge of which is taken to be 100 units. It follows that the altitude of each triangle face $= 86 \cdot 6$.

The outside values are given in the manner most convenient for the construction of a model or drawing. If any point lies in an edge of the fundamental tetrahedron, its two co-ordinates (finite) are in the ratio in which the edge is divided. So I have taken the sum of the two co-ordinates for a point on an edge $= 100$. Recalling the fact that, for an equilateral triangle, the sum of the co-ordinates for a point in the plane of the triangle is equal to the altitude of the triangle, I have so chosen the co-ordinates of a point lying in a face of the tetrahedron, that their sum is equal to $86 \cdot 6$; and in this case, the three co-ordinates denote the perpendicular distances from the sides of the triangle.

Some little ingenuity had to be exercised in constructing the model, especially in deciding which faces of the fundamental tetrahedron to dispense with, in order to leave the model open to view. On laying down the points, it appeared that the model might be constructed by using only the planes of ABD and BCD, since the planes ABC and ADC are intersected by lines of the double six at no points not lying on the edges BC and DC. This statement will be clear from an inspection of Plate 1, a perspective drawing of the configuration, made exactly according to the directions given above.

CHAPTER VI

THE CONSTRUCTION OF THE CONFIGURATIONS OF THE STRAIGHT LINES UPON THE TWENTY-ONE TYPES OF THE CUBIC SURFACE

20. Division of the Cubic Surface into Types.

Schläfli* first conceived the idea of a division of the cubic surface into species in reference to the reality of the straight lines lying upon the surface. He later published an extensive memoir on the subject†, which served as a basis for Cayley's exhaustive *Memoir on Cubic Surfaces*‡. The general surface of the third order falls into only five different types§, in reference to the reality of the twenty-seven lines. These are shown in the following table:

Type	No. Real Planes	No. Real Lines
1	45	27
2	15	15
3	5	7
4	7	3
5	13	3

This conclusion follows from the fact that every surface of the third order (real, general) may be generated by means of two triheders which present one or the other of the following three cases: (1) the

* *Quart. Journ.* Vol. II. (1858), pp. 55–65, 110–120.
† *Philos. Trans. Royal Soc.* Vol. CLIII. (1863), pp. 193–241.
‡ *Philos. Trans. Royal Soc.* Vol. CLIX. (1869), pp. 231–326.
§ Cremona, *Crelle's Journ.* Vol. LXVIII. (1868), pp. 1–133.

triheders are formed by six real planes; (2) one triheder is wholly real, while the other is formed by one real and two conjugate imaginary planes; (3) each triheder has one real and two conjugate imaginary planes.

The division according to the nature of the singularities, however, is the division with which we shall be concerned, in particular, in the sequel (cf. Schläfli's second paper, l.c., and Cayley's *Memoir on Cubic Surfaces*).

21. On the Construction of Models of the Twenty-seven Lines upon the Cubic Surface.

For the first time in 1869, Dr Christian Wiener* constructed a model of the twenty-seven lines upon the general cubic surface without singularities (cf. historical summary). Sylvester† and Salmon‡ had each given the same method of construction; but this was a "pure geometric" construction, without indication as to how it was to be carried out in making a model (thread, wire, plaster, or otherwise) of the configuration.

In 1882, Percival Frost§ gave a full description of the method by means of which he was enabled to make a thread model of the twenty-seven lines. Theoretically his method is based upon the analytical investigation of Schläfli‖, in his original paper upon the twenty-seven lines. Frost determined the equations of the twenty-seven lines; and then, giving numerical values to the constants involved, he calculated in cartesian co-ordinates the co-ordinates of the one hundred and thirty-five points of intersection of the twenty-seven lines. Much care had to be exercised in the choice of numerical values for the constants, in order that all the lines might fall within reach; that the triangles might appear fairly spread out; and that coincidence or parallelism among the lines might be avoided. The method was tremendously laborious; and even after the model was constructed by Frost, it was imperfect or rather incomplete, several of the lines falling entirely out of reach.

Blythe¶ has given a purely geometrical method for constructing a model of the lines upon the cubic surface without singularities, and has

* Cayley, *Trans. Camb. Philos. Soc.* Vol. XII. Part I. (1873), pp. 366–383.

† *Comptes Rendus*, Vol. LII. (1861), pp. 977–980.

‡ *Geometry of Three Dimensions*, 4th edition, p. 500.

§ *Quart. Journ.* Vol. XVIII. (1882), pp. 89–96.

‖ *Quart. Journ.*, l.c.

¶ *Quart. Journ.* Vol. XXIX. (1898), pp. 206–223.

briefly sketched the variations for the types of the surface possessed of singularities.

This method is based initially upon the fact that the general equation of the cubic surface may be put into the form

$$\alpha\beta\gamma = K\delta S,$$

where α, β, γ, δ are of the first degree, S is of the second degree, in the variables, and K a constant*. Taking a series of planes through some straight line lying wholly upon this surface, from the form of the equation the conclusion is drawn that the series of conic sections thus obtained cuts the line in pairs of points which are a range in involution. It is shown (l.c.) that a cubic surface may be determined by straight lines l, m, n; L, M, N; P, Q, R, S, conditioned as follows:

l, m, n are co-planar
L, M, N " "
L, P, Q " "
L, R, S " "

provided M, P, R meet m, and N, Q, S meet n: where M, N; P, Q; R, S cut L in points, which are a range in involution.

While these investigations are very interesting, the construction is not taken in all its generality; in a number of instances a special investigation is required, necessitating a marked modification of the process for the general case.

More recently, Blythe has considered the subject again in a brief paper†, this time giving the construction for the twenty-seven lines upon only the cubic surface without singularities, again employing the notion of five pairs of points of an involution. His researches are recorded in full in his book, *On Models of Cubic Surfaces* (Cambridge University Press).

The papers of Korteweg‡ and De Vries§ also deserve special mention.

22. A Uniform Method for the Construction of Models.

In the present article, there is detailed a uniform method of representing the lines on each type of the cubic surface by means of models or perspective drawings. I have actually constructed, to scale, a

* Cayley, *Coll. Math. Papers*, Vol. I. No. 76.

† *Quart. Journ.* Vol. XXXIV. No. 1 (1902), pp. 73–74.

‡ *Nieuw Archief voor Wiskunde*, Amsterdam, Vol. XX. (1893).

§ *Archives Néerlandaises des sciences exactes et naturelles*, Haarlem, Sér. 2, Vol. VI. (1901).

graphical representation of the lines in all twenty-one types of the cubic surface.

I have made use of the materials found in Cayley's *Collected Math. Papers*, Vol. VI. No. 412, employing the equations of the lines and triple tangent planes of the various surfaces, expressed in the canonical form. This has been done in every case save that of the surface without singularities, the most general case. The problem in this case is greatly simplified by choosing an equation for the surface with highly specialized coefficients. All these equations are expressed in quadriplanar co-ordinates. In the drawings or models, the lines appear not only in proper relation to each other, but also in correct position with respect to the edges of the fundamental tetrahedron. The notation employed here for the lines and planes, and also for each type of the cubic (with the exception of the most general type) is identical with that found in the article last referred to. Although the division depending upon the singularities gives rise to twenty-three types of the cubic surface, two of these are scrolls, in which there is no question of the twenty-seven lines.

A few words of explanation will serve to make the method clear. For each type of the cubic surface, I have chosen, by trial, such numerical values for the constants as will show the entire configuration in a comparatively limited space. In each case, I have made out a table of numerical co-ordinates for the points where each line meets two faces of the fundamental tetrahedron, and these points, of course, fix each line in position. A fact of essential simplicity is that, in almost every case, only two planes of the fundamental tetrahedron are used, these leaving the model wholly open to view.

To illustrate the process, take for example the general type of the cubic surface. The lines 1 and $2'$ coincide with the edges BC and CD, respectively, of the fundamental tetrahedron $ABCD$. Consider now the line 12, which meets 1 and $2'$. The line 12 lies in the plane $12.34.56$; hence to find where the line 12 meets the line 1, we make $x=0$, $w=0$ in the equation of the plane $12.34.56$. Similarly, to find where the line 12 meets the edge CD ($x=0$, $y=0$), which is the line $2'$, make $x=0$, $y=0$ in the equation of the plane $12.34.56$. Planes passing through the line 12, other than the plane $12.34.56$, might have been used; and the simplest form has always been used.

In the case of a line such as 45, which meets the tetrahedron in only one edge AB, it remains to find where the line 45 meets the plane BCD ($x=0$). Here we consider two planes, $45'$ and $54'$, which intersect in the line 45. Making $x=0$ in the equations of each one of these planes, we find their lines of intersection with the plane BCD; and these lines intersect where 45 meets the plane BCD. Similar reasoning holds for a line which meets no edge of the fundamental tetrahedron.

The values given in the tables of co-ordinates have been carefully checked in every case. Using some other planes than the two used in determining the co-ordinates of the point in question, it has been verified that the co-ordinates actually do satisfy the equations of these other planes upon which the point should lie. Just as in the construction of the model of the double six, I have for convenience taken x, y, z, w as the perpendicular distances of the current point from the faces of a regular tetrahedron, whose edge is taken to be 100 units; in consequence, the altitude of each face is equal to 86·6. Accordingly, for the values on the right in the tables, the following checks were used: if the point lies on an edge, the sum of the co-ordinates equals 100; if the point is not on an edge, the sum of the co-ordinates equals 86·6.

23. Concerning the Graphical Representation of the Configurations of the Lines on the Cubic Surface.

In making accurate drawings, to scale, of the various configurations, there were a number of difficulties to be overcome. Five separate scales had to be brought into play.

First, a tetrahedron was drawn, attention being paid to the requirements of perspective. It will be recalled that a point which lies on an edge has, for co-ordinates, the distances to the extremities of the edge. Moreover, when a point lies in a face of the fundamental tetrahedron, the perpendicular distances to the sides of that triangle are the co-ordinates of the point. The tetrahedron is taken to be regular, with each edge equal to 100 units in length.

I shall describe the method for making the diagram of the configuration of the twenty-seven lines on the general surface of the third degree; but the principles employed are the same for all the graphic representations. First are made four separate scales, one for each of the edges AB, BC, CD, DA, of lengths as laid down on the diagram; each scale is divided into 100 equal parts. Next, using the proper scale for each edge, I laid down the position of each point where the four edges AB, BC, CD, DA, i.e. the lines 4′, 1, 2′ and 3 respectively, are met by the remaining lines.

The following method was employed for determining with accuracy the positions of the points, in the faces ABD, BCD, not lying on an edge of the tetrahedron. I first drew on two separate sheets, one on each sheet, two equilateral triangles, representing the two faces ABD, BCD. These triangles were constructed equal in size. Next was constructed a new scale, of length equal to the side of one of the triangles, and divided into 100 equal parts. I next located on these two sheets, to the new scale, each point, not on an edge, which lies in one of these faces. I next determine its position by lines drawn parallel to the sides of the triangle, two such parallels in each case being sufficient to fix the position of the point; and then find the lengths of the parallels according to the scale of the triangle.

Now, returning to the original perspective representation of the tetrahedron, I locate each point by like parallels, using for each parallel the scale of that edge, parallel to which the line is drawn, as originally chosen.

An immediate check on the accuracy of the construction is afforded by the fact that all the points, so constructed, that lie in any one face, are collinear. This graphical check is likewise a rough check on the accuracy of the numerical computations for the co-ordinates of the points.

24. Species I = 12.

The equation of the surface is taken to be (§ 16)

$$\left(\frac{x}{x_2}+\frac{y}{y_2}+\frac{z}{z_2}+\frac{w}{w_2}\right)\left(\frac{xz}{x_1z_1}-\frac{yw}{y_1w_1}\right)$$
$$-k\left(\frac{x}{x_1}+\frac{y}{y_1}+\frac{z}{z_1}+\frac{w}{w_1}\right)\left(\frac{xz}{x_2z_2}-\frac{yw}{y_2w_2}\right)=0.$$

The equations of the forty-five triple tangent planes are given in § 16, and these determine, in pairs, the equations of the twenty-seven lines upon the surface.

The values assigned for the constants are as follows:

$$x_1=3,\ y_1=-4,\ z_1=5,\ w_1=-6;$$
$$x_2=z_2=1;\ y_2=w_2=-1;\ k=-\frac{3}{16}.$$

Below is given a table of numerical co-ordinates which serve to fix definitely the lines in position. It should be observed that the fact of essential simplicity in connection with the construction of this model is the employment of only two planes, viz. the planes of the two faces *ABD*, *BCD* of the fundamental tetrahedron *ABCD*, whereby the model is left entirely open to view. The intersection table of the twenty-seven lines (§ 11), considered in connection with the diagram of the present configuration, shows immediately which lines do, and which lines do not, intersect. Actually, the notation attached to the lines is entirely self-explanatory on this point (Plate 2).

	x	y	z	w	Co-ordinates, for edge=100
1 is line *BC*	0			0	
2′ „ „ *CD*	0	0			
3 „ „ *DA*		0	0		
4′ „ „ *AB*			0	0	
5 meets *CD*			5	6	$z=45{\cdot}5$, $w=54{\cdot}5$
„ *AB*	3	4			$x=42{\cdot}9$, $y=57{\cdot}1$
6 „ *CD*			1	1	$z=50$, $w=50$
„ *AB*	1	1			$x=50$, $y=50$
6′ „ *BC*		4	5		$y=44{\cdot}4$, $z=55{\cdot}6$
„ *AD*	3			6	$x=33{\cdot}3$, $w=66{\cdot}7$
5′ „ *BC*		1	1		$y=50$, $z=50$
„ *AD*	1			1	$x=50$, $w=50$
1′ „ *AD*	25			34	$x=42{\cdot}4$, $w=57{\cdot}6$
„ *BCD*		−100	170	306	$y=-23$, $z=39{\cdot}1$, $w=70{\cdot}4$
3′ „ *BC*		28	31		$y=47{\cdot}5$, $z=52{\cdot}5$
„ *ABD*	84	252		−186	$x=48{\cdot}5$, $y=145{\cdot}5$, $w=-107{\cdot}4$
2 „ *AB*	25	28			$x=47{\cdot}2$, $y=52{\cdot}8$
„ *BCD*		252	420	150	$y=26{\cdot}6$, $z=44{\cdot}3$, $w=15{\cdot}7$
4 *CD*			31	34	$z=47{\cdot}7$, $w=52{\cdot}3$
„ *ABD*	102	372		−306	$x=52{\cdot}6$, $y=191{\cdot}7$, $w=-157{\cdot}7$
„ *ABC*	102	372	279		$x=11{\cdot}7$, $y=42{\cdot}8$, $z=32{\cdot}1$
12 „ *BC*		76	85		$y=47{\cdot}2$, $z=52{\cdot}8$
„ *CD*			35	38	$z=48$, $w=52$

	x	y	z	w	Co-ordinates, for edge=100
13 meets BC		−10	17		$y=-143,\ z=243$
„ AD	−14			31	$x=-82{\cdot}3,\ w=182{\cdot}3$
14 „ BC		100	109		$y=47{\cdot}8,\ z=52{\cdot}2$
„ AB	109	124			$x=46{\cdot}8,\ y=53{\cdot}2$
15 „ BC		20	17		$y=54,\ z=46$
„ ABD	17	6		17	$x=36{\cdot}8,\ y=13,\ w=36{\cdot}8$
16 „ BC		25	34		$y=42{\cdot}4,\ z=57{\cdot}6$
„ ABD	17	−8		34	$x=34{\cdot}2,\ y=-16{\cdot}1,\ w=68{\cdot}5$
23 „ CD			47	50	$z=48{\cdot}5,\ w=51{\cdot}5$
„ AD	47			62	$x=43{\cdot}1,\ w=56{\cdot}9$
24 „ CD			14	5	$z=73{\cdot}7,\ w=26{\cdot}3$
„ AB	17	62			$x=21{\cdot}5,\ y=78{\cdot}5$
25 „ CD			28	25	$z=52{\cdot}8,\ w=47{\cdot}2$
„ ABD	21	28		−4	$x=40{\cdot}4,\ y=53{\cdot}9,\ w=-7{\cdot}7$
26 „ CD			7	10	$z=41{\cdot}2,\ w=58{\cdot}8$
„ ABD	14	14		3	$x=39{\cdot}1,\ y=39{\cdot}1,\ w=8{\cdot}4$
34 „ AD	7			10	$x=41{\cdot}2,\ w=58{\cdot}8$
„ AB	17	20			$x=45{\cdot}9,\ y=54{\cdot}1$
35 „ AD	35			62	$x=36{\cdot}1,\ w=63{\cdot}9$
„ BCD		70	70	−9	$y=46{\cdot}3,\ z=46{\cdot}3,\ w=-6$
36 „ AD	28			31	$x=47{\cdot}5,\ w=52{\cdot}5$
„ BCD		28	35	4	$y=36{\cdot}2,\ z=45{\cdot}2,\ w=5{\cdot}2$
45 „ AB	34	31			$x=52{\cdot}3,\ y=47{\cdot}7$
„ BCD		−24	255	306	$y=-3{\cdot}8,\ z=41{\cdot}1,\ w=49{\cdot}3$
46 „ AB	85	124			$x=40{\cdot}7,\ y=59{\cdot}3$
„ BCD		6	85	85	$y=3,\ z=41{\cdot}8,\ w=41{\cdot}8$
56 „ ABD	1	2		−1	$x=43{\cdot}3,\ y=86{\cdot}6,\ w=-43{\cdot}3$
„ BCD		2	5	3	$y=17{\cdot}3,\ z=43{\cdot}3,\ w=26$

On the basis of the above data, in connection with the results of § 10, it is of course a mere matter of detail to construct all of the thirty-six double sixes belonging to the cubic surface.

25. Species $\mathrm{II} = 12 - C_2$.

The equation of the surface is:

$$w\,(a,\ b,\ c,\ f,\ g,\ h \between x,\ y,\ z)^2 + 2Kxyz = 0.$$

Let us write this equation in the form

$$w\left(1,\ 1,\ 1,\ l+\frac{1}{l},\ m+\frac{1}{m},\ n+\frac{1}{n} \between x,\ y,\ z\right)^2 + \frac{\alpha\beta\gamma\delta}{p}\,xyz = 0,$$

where for brevity

$$\begin{aligned} \alpha &= mn - l, \\ \beta &= nl - m, \\ \gamma &= lm - n, \\ \delta &= lmn - 1, \\ p &= lmn. \end{aligned}$$

If we take $x=0$ as the equation of the plane [12], $y=0$ as that of the plane [34], $z=0$ as that of the plane [56], then the equations of the thirty distinct tangent planes to the cubic surface may be written down immediately; from these are obtained the equations of the twenty-one distinct lines upon the surface in the following forms:

(1) : $x=0,\ y+lz=0,$

(3) : $y=0,\ z+mx=0,$

(5) : $z=0,\ x+ny=0,$

(2) : $x=0,\ y+l^{-1}z=0,$

(4) : $y=0,\ z+m^{-1}x=0,$

(6) : $z=0,\ x+n^{-1}y=0,$

(45) : $x+ny+mz=0,\ x+\beta\gamma w=0,$

(16) : $y+lz+nx=0,\ y+\gamma\alpha w=0,$

(23) : $z+mx+ly=0,\ z+\alpha\beta w=0,$

(46) : $x+n^{-1}y+mz=0,\ x-\alpha\delta w=0,$

(26) : $y+l^{-1}z+nx=0,\ y-\beta\delta w=0,$

(24) : $z+m^{-1}x+ly=0,\ z-\gamma\delta w=0,$

(35) : $x+ny+m^{-1}z=0,\ x-\alpha\delta w=0,$

(15) : $y+lz+n^{-1}x=0,\ y-\beta\delta w=0,$

(13) : $z+mx+l^{-1}y=0,\ z-\gamma\delta w=0,$

(36) : $x+n^{-1}y+m^{-1}z=0,\ x+\beta\gamma w=0,$

$$(25): \quad y + l^{-1}z + n^{-1}x = 0, \; y + \gamma a w = 0,$$
$$(14): \quad z + m^{-1}x + l^{-1}y = 0, \; z + \alpha\beta w = 0,$$
$$(12): \quad x = 0, \; w = 0,$$
$$(34): \quad y = 0, \; w = 0,$$
$$(56): \quad z = 0, \; w = 0.$$

Note that p does not appear in the equations of any of the twenty-one lines.

	x	y	z	w	Co-ordinates, for edge = 100
45 meets BC		4	−9		$y=80$, $z=-180$
„ ACD	−187		561	288	$x=-24{\cdot}4$, $z=73{\cdot}4$, $w=37{\cdot}6$
16 „ AC	2		3		$x=40$, $z=60$
„ ABD	−44	33		48	$x=-103$, $y=77{\cdot}2$, $w=112{\cdot}4$
23 „ AB	3	2			$x=60$, $y=40$
„ BCD		34	17	32	$y=35{\cdot}5$, $z=17{\cdot}7$, $w=33{\cdot}4$
46 „ BC		−1	4		$y=-33{\cdot}3$, $z=133{\cdot}3$
„ ACD	−27		81	32	$x=-27{\cdot}2$, $z=81{\cdot}6$, $w=32{\cdot}2$
26 „ AC	8		3		$x=72{\cdot}8$, $z=27{\cdot}2$
„ BCD		102	51	128	$y=31{\cdot}4$, $z=15{\cdot}7$, $w=39{\cdot}5$
24 „ AB	1	6			$x=14{\cdot}3$, $y=85{\cdot}7$
„ ACD	−11		33	32	$x=-17{\cdot}6$, $z=52{\cdot}9$, $w=51{\cdot}3$
35 „ BC		4	−1		$y=133{\cdot}3$, $z=-33{\cdot}3$
„ ABD	−27	36		32	$x=-57$, $y=76$, $w=67{\cdot}6$
15 „ AC	3		8		$x=27{\cdot}2$, $z=72{\cdot}8$
„ BCD		51	102	64	$y=20{\cdot}3$, $z=40{\cdot}6$, $w=25{\cdot}6$
13 „ AB	6	1			$x=85{\cdot}7$, $y=14{\cdot}3$
„ BCD		33	66	64	$y=17{\cdot}5$, $z=35{\cdot}1$, $w=34$
36 „ BC		9	−4		$y=180$, $z=-80$
„ ABD	−748	561		1152	$x=-67{\cdot}1$, $y=50{\cdot}3$, $w=103{\cdot}4$
25 „ AC	3		2		$x=60$, $z=40$
„ BCD		22	11	32	$y=29{\cdot}3$, $z=14{\cdot}7$, $w=42{\cdot}6$
14 „ AB	2	3			$x=40$, $y=60$
„ BCD		17	34	64	$y=12{\cdot}8$, $z=25{\cdot}6$, $w=48{\cdot}2$

In these, let us set

$$l = -\frac{1}{2}, \quad m = +\frac{1}{3}, \quad n = +\frac{3}{4}.$$

Consequently

$$\alpha = \frac{3}{4}, \quad \beta = -\frac{17}{24}, \quad \gamma = -\frac{11}{12}, \quad \delta = -\frac{9}{8}.$$

Calculating the co-ordinates of the points where each line meets two faces of the tetrahedron $ABCD$ (except in the case of the line 12, 34, 56, lying wholly in the face ABC, and the lines 1, 2, 3, 4, 5, 6, for each one of which, since they lie by pairs in the faces ABD, BCD, ACD, only one point has to be calculated), we obtain the results as tabulated above (p. 67).

It is to be observed that the system of lines and planes is at once deduced from that belonging to the general equation of the cubic surface, by supposing that in the primitive double six the corresponding lines 1 and 1′, 2 and 2′, etc. severally coincide.

In the present numbering, the lines 1, 2, 3, 4, 5, 6 pass through a conical node at D on the surface, i.e. they lie not only in the cubic surface, but also in a quadric cone of vertex D (Plate 3).

26. Species III $= 12 - B_3$.

The equation of the surface is

$$2w(x+y+z)(lx+my+nz)+2kxyz=0,$$

where we set

$$\lambda, \ \mu, \ \nu \equiv m-n, \ n-l, \ l-m, \text{ respectively.}$$

Let us take $\quad l = -\frac{1}{2}, \quad m = +\frac{1}{3}, \quad n = \frac{3}{4}.$

Consequently $\quad \lambda = -\frac{5}{12}, \quad \mu = \frac{5}{4}, \quad \nu = -\frac{5}{6}.$

The equations of the fifteen distinct lines upon the surface take the form :

$$\begin{aligned}
(1):&\quad x=0, \ y+z=0,\\
(2):&\quad y=0, \ z+x=0,\\
(3):&\quad z=0, \ x+y=0,\\
(4):&\quad x=0, \ 4y+9z=0,\\
(5):&\quad y=0, \ 3z-2x=0,\\
(6):&\quad z=0, \ -3x+2y=0,\\
(14):&\quad x=0, \ w=0,\\
(25):&\quad y=0, \ w=0,\\
(36):&\quad z=0, \ w=0,
\end{aligned}$$

$$(15): \quad 2x - 3y - 3z = 0, \ 8w - 5z = 0,$$
$$(16): \quad 3x - 2y - 2z = 0, \ 12w + 5y = 0,$$
$$(26): \quad 3x - 2y + 3z = 0, \ 24w + 5x = 0,$$
$$(24): \quad 9x + 4y + 9z = 0, \ 8w - 5z = 0,$$
$$(34): \quad 4x + 4y + 9z = 0, \ 12w + 5y = 0,$$
$$(35): \quad 2x + 2y - 3z = 0, \ 24w + 5x = 0.$$

The plane of ABC ($w = 0$) was chosen for the base of the tetrahedron, since it contains three of the lines. The lines 14, 25, 36 are the edges BC, AC, AB, respectively, of the tetrahedron. The lines 1, 2, and 3 pass through the point D and bisect the angle D in the planes of BCD, CDA, ABD respectively. The data for the remaining lines are given in the following table :

	x	y	z	w	Co-ordinates, for edge = 100
4 passes through	D in plane BCD				
meets BC		9	−4		
5 passes through	D in plane ACD				
meets AC	3		2		
6 passes through	D in plane ABD				
meets AB	2	3			
15 meets AB	3	2			
,, ACD	12		8	5	$x = 41{\cdot}6$, $z = 27{\cdot}7$, $w = 17{\cdot}3$
16 ,, AC	2		3		
,, ABD	8	12		−5	$x = 46{\cdot}2$, $y = 69{\cdot}3$, $w = -28{\cdot}9$
26 ,, BC		3	2		
,, ABD	24	36		−5	$x = 37{\cdot}8$, $y = 56{\cdot}7$, $w = -7{\cdot}9$
24 ,, AB	−4	9			
,, ACD	−8		8	5	$x = -138{\cdot}6$, $z = 138{\cdot}6$, $w = 86{\cdot}6$
34 ,, AC	9		−4		
,, ABD	12	−12		5	$x = 207{\cdot}8$, $y = -207{\cdot}8$, $w = 86{\cdot}6$
35 ,, BC		3	2		
,, ACD	24		16	−5	$x = 59{\cdot}4$, $z = 39{\cdot}6$, $w = -12{\cdot}4$

The system of lines and planes for this surface is deduced immediately from that belonging to Species II by supposing the tangent cone to reduce itself to the pair of biplanes; three of the planes of Species II come to coincide with one biplane, three of them with the other. The line of intersection of the biplanes is called the edge.

In the present case, the point D is a binode on the surface. The edge is not a line on the surface (Plate 4).

27. Species IV = 12 − 2C_2.

The equation of the surface is

$$wxz + y^2(\gamma z + \delta w) + (a, b, c, d \between x, y)^3 = 0,$$

where we set

$$x(a, b, c, d \between x, y)^3 - \gamma\delta y^4 = \frac{-\gamma\delta}{f_1 f_2 f_3 f_4}(x - f_1 y)(x - f_2 y)(x - f_3 y)(x - f_4 y).$$

Let us take

$$f_1 = 1,\ f_2 = \frac{1}{2},\ f_3 = \frac{1}{3},\ f_4 = \frac{1}{4};\ \gamma = -\frac{1}{3},\ \delta = -\frac{2}{5};$$

$$d = \gamma\delta\left(\frac{1}{f_1} + \frac{1}{f_2} + \frac{1}{f_3} + \frac{1}{f_4}\right) = \frac{4}{3}.$$

The equations of the sixteen distinct lines upon the surface take the forms:

$$\begin{array}{rl}
(0): & x = 0,\ y = 0, \\
(5): & x = 0,\ 20y - 5z - 6w = 0, \\
(1): & x - y = 0,\ 2y - 5z = 0, \\
(2): & 2x - y = 0,\ 4y - 5z = 0, \\
(3): & 3x - y = 0,\ 6y - 5z = 0, \\
(4): & 4x - y = 0,\ 8y - 5z = 0, \\
(1'): & x - y = 0,\ y - 3w = 0, \\
(2'): & 2x - y = 0,\ 2y - 3w = 0, \\
(3'): & 3x - y = 0,\ y - w = 0, \\
(4'): & 4x - y = 0,\ 4y - 3w = 0, \\
(12\,.\,3'4'): & 4x - 6y + 5z = 0,\ 12x - 7y + 3w = 0, \\
(13\,.\,2'4'): & 6x - 8y + 5z = 0,\ 8x - 6y + 3w = 0, \\
(14\,.\,2'3'): & 8x - 10y + 5z = 0,\ 6x - 5y + 3w = 0, \\
(23\,.\,1'4'): & 12x - 10y + 5z = 0,\ 4x - 5y + 3w = 0, \\
(24\,.\,1'3'): & 16x - 12y + 5z = 0,\ 3x - 4y + 3w = 0, \\
(34\,.\,1'2'): & 24x - 14y + 5z = 0,\ 2x - 3y + 3w = 0.
\end{array}$$

The plane of BCD ($x = 0$) was chosen for the base of the tetrahedron, since it contains two of the lines. The line (0) is the edge CD of the tetrahedron. The data for the remaining lines are given in the following table:

	x	y	z	w	Co-ordinates, for edge = 100
0 is line CD					
5 meets BD		3		10	$y=23,\ w=77$
„ BC		1	4		$y=20,\ z=80$
1 passes through D meets ABC	5	5	2		$x=36{\cdot}1,\ y=36{\cdot}1,\ z=14{\cdot}4$
2 passes through D meets ABC	5	10	8		$x=18{\cdot}8,\ y=37{\cdot}7,\ z=30{\cdot}1$
3 passes through D meets ABC	5	15	18		$x=11{\cdot}4,\ y=34{\cdot}2,\ z=41$
4 passes through D meets ABC	5	20	32		$x=7{\cdot}6,\ y=30{\cdot}4,\ z=48{\cdot}6$
1′ passes through C meets ABD	3	3		1	$x=37{\cdot}1,\ y=37{\cdot}1,\ w=12{\cdot}4$
2′ passes through C meets ABD	3	6		4	$x=20,\ y=40,\ w=26{\cdot}6$
3′ passes through C meets ABD	1	3		3	$x=12{\cdot}4,\ y=37{\cdot}1,\ w=37{\cdot}1$
4′ passes through C meets ABD	3	12		16	$x=8{\cdot}4,\ y=33{\cdot}5,\ w=44{\cdot}7$
12.3′4′ meets BCD		630	756	1470	$y=19{\cdot}1,\ z=22{\cdot}9,\ w=44{\cdot}6$
„ ABC	8085	13860	10164		$x=21{\cdot}8,\ y=37{\cdot}4,\ z=27{\cdot}4$
13.2′4′ „ BCD		5	8	10	$y=18{\cdot}8,\ z=30{\cdot}1,\ w=37{\cdot}7$
„ ABC	15	20	14		$x=26{\cdot}5,\ y=35{\cdot}4,\ z=24{\cdot}7$
14.2′3′ „ BCD		3	6	5	$y=18{\cdot}6,\ z=37{\cdot}1,\ w=30{\cdot}9$
„ ABC	5	6	4		$x=28{\cdot}9,\ y=34{\cdot}6,\ z=23{\cdot}1$
23.1′4′ „ BCD		3	6	5	$y=18{\cdot}6,\ z=37{\cdot}1,\ w=30{\cdot}9$
„ ABD	15	18		10	$x=30{\cdot}2,\ y=36{\cdot}3\ \ w=20{\cdot}1$
24.1′3′ „ BCD		15	36	20	$y=18{\cdot}3,\ z=43{\cdot}9,\ w=24{\cdot}4$
„ ABD	9	12		7	$x=27{\cdot}8,\ y=37{\cdot}1,\ w=21{\cdot}6$
34.1′2′ „ BCD		5	14	5	$y=18{\cdot}1,\ z=50{\cdot}5,\ w=18{\cdot}1$
„ ABD	9702	16632		10164	$x=23,\ \ y=39{\cdot}5,\ w=24{\cdot}1$

In the present case, there are two conic nodes, at C and D respectively. The rays 1, 2, 3, 4 and 1′, 2′, 3′, 4′ pass through the two nodes D and C, respectively (Plate 5).

28. Species $V = 12 - B_4$.

The equation of the surface is

$$wxz + (x+z)(y^2 - ax^2 - bz^2) = 0.$$

The equations of the ten distinct lines upon the surface are as follows :

$$\begin{aligned}
(3):&\quad x=0,\ z=0,\\
(4):&\quad x+z=0,\ w=0,\\
(1):&\quad x=0,\ y-\sqrt{b}\,z=0,\\
(2):&\quad x=0,\ y+\sqrt{b}\,z=0,\\
(1'):&\quad z=0,\ -\sqrt{a}\,x+y=0,\\
(2'):&\quad z=0,\ \sqrt{a}\,x+y=0,\\
(11'):&\quad -\sqrt{a}\,x+y-\sqrt{b}\,z=0,\ \sqrt{ab}\,(x+z)+w=0,\\
(12'):&\quad \sqrt{a}\,x+y-\sqrt{b}\,z=0,\ -2\sqrt{ab}\,(x+z)+w=0,\\
(21'):&\quad -\sqrt{a}\,x+y+\sqrt{b}\,z=0,\ -2\sqrt{ab}\,(x+z)+w=0,\\
(22'):&\quad \sqrt{a}\,x+y+\sqrt{b}\,z=0,\ \sqrt{ab}\,(x+z)+w=0.
\end{aligned}$$

The numerical values chosen for the constants are

$$a=\frac{4}{9},\ b=\frac{9}{25}.$$

The plane of ABC ($z=0$) was selected for the base of the tetrahedron, since it contains three of the lines. The line (3) is the edge BD, the line (4) passes through A and is parallel to the edge BC. The data for the remaining lines are given in the table on page 73.

There is a binode at D, and the edge is torsal, i.e. the surface is touched along the edge by a plane (Plate 6).

29. Species $VI = 12 - B_3 - C_2$.

The equation of the surface is

$$wxz + y^2z + (a,\ b,\ c,\ d\between x,\ y)^3 = 0,$$

where we set

$$(a,\ b,\ c,\ d\between x,\ y)^3 \equiv -d\,(\theta_2 x - y)(\theta_3 x - y)(\theta_4 x - y).$$

The numerical values assigned for the constants are as follows :

$$\theta_2=\frac{2}{3},\ \theta_3=\frac{4}{5},\ \theta_4=\frac{3}{4};\ d=1.$$

	x	y	z	w	Co-ordinates, for edge=100
1 passes through	D in plane BCD				
meets BC		3	5		
2 passes through	D in plane BCD				
meets BC		−3	5		
1′ passes through	D in plane ABD				
meets AB	3	2			
2′ passes through	D in plane ABD				
meets AB	3	−2			
11′ meets ABD	15	10		−6	$x=68{\cdot}3$, $y=45{\cdot}6$, $w=-27{\cdot}3$
" BCD		3	5	−2	$y=43{\cdot}3$, $z=72{\cdot}2$, $w=-28{\cdot}9$
12′ " ABD	15	−10		12	$x=76{\cdot}4$, $y=-50{\cdot}9$, $w=61{\cdot}1$
" BCD		3	5	4	$y=21{\cdot}6$, $z=36{\cdot}1$, $w=28{\cdot}9$
21′ " ABD	15	10		12	$x=35{\cdot}1$, $y=23{\cdot}4$, $w=28{\cdot}1$
" BCD		−3	5	4	$y=-43{\cdot}3$, $z=72{\cdot}2$, $w=57{\cdot}7$
22′ " ABD	−15	10		6	This line is not shown on the drawing, since it falls off the sheet
" BCD		3	−5	2	

The equations of the eleven distinct lines upon the surface are as follows :

$$
\begin{aligned}
(0):&\quad x=0,\ y=0,\\
(1):&\quad x=0,\ y+z=0,\\
(2):&\quad 2x-3y=0,\ z=0,\\
(3):&\quad 4x-5y=0,\ z=0,\\
(4):&\quad 3x-4y=0,\ z=0,\\
(2'):&\quad 2x-3y=0,\ 4x+9w=0,\\
(3'):&\quad 4x-5y=0,\ 16x+25w=0,\\
(4'):&\quad 3x-4y=0,\ 9x+16w=0,\\
(12\,.\,3'4'):&\quad 2x-3y-3z=0,\ 12x-31y-20w=0,\\
(13\,.\,2'4'):&\quad 4x-5y-5z=0,\ 6x-17y-12w=0,\\
(14\,.\,2'3'):&\quad 3x-4y-4z=0,\ 8x-22y-15w=0.
\end{aligned}
$$

The plane of ABD ($z=0$) was chosen for the base of the tetrahedron, since it contains three of the lines. The line (0) is the edge CD of the tetrahedron ; the line (1) passes through D and is parallel to the edge BC. The data for the remaining lines are given in the following table:

	x	y	z	w	Co-ordinates, for edge = 100
2 passes through	D in plane ABD				
meets AB	3	2			
3 passes through	D in plane ABD				
meets AB	5	4			
4 passes through	D in plane ABD				
meets AB	4	3			
2′ passes through C meets ABD	9	6		−4	$x=70{\cdot}9$, $y=47{\cdot}2$, $w=-31{\cdot}5$
3′ passes through C meets ABD	25	20		−16	$x=74{\cdot}7$, $y=59{\cdot}7$, $w=-47{\cdot}8$
4′ passes through C meets ABD	16	12		−9	$x=72{\cdot}9$, $y=54{\cdot}7$, $w=-41$
12.3′4′ meets ABD	30	20		−13	$x=70{\cdot}2$, $y=46{\cdot}8$, $w=-30{\cdot}4$
,, ACD	15		10	9	$x=38{\cdot}2$, $z=25{\cdot}5$, $w=22{\cdot}9$
13.2′4′ ,, ABD	30	24		−19	$x=74{\cdot}2$, $y=59{\cdot}4$, $w=-47$
,, ACD	10		8	5	$x=37{\cdot}7$, $z=30{\cdot}1$, $w=18{\cdot}8$
14.2′3′ ,, ABD	60	45		−34	$x=73{\cdot}2$, $y=54{\cdot}9$, $w=-41{\cdot}5$
,, ACD	60		45	32	$x=37{\cdot}9$, $z=28{\cdot}5$, $w=20{\cdot}2$

Here there is a binode at D, a conic node at C. The axis joining the two nodes is a line on the surface (Plate 7).

30. Species VII = 12 − B_5.

The equation of the surface is

$$wxz + y^2z + yx^2 - z^3 = 0.$$

The equations of the six distinct lines upon the surface are as follows:

$$(0):\ x=0,\ z=0,$$
$$(1):\ y=0,\ z=0,$$
$$(2'):\ x=0,\ y+z=0,$$
$$(3'):\ x=0,\ y-z=0,$$
$$(12'):\ x-w=0,\ y+z=0,$$
$$(13'):\ x+w=0,\ y-z=0.$$

The plane of BCD ($x=0$) was chosen for the base of the tetrahedron, since it contains three of the lines. The lines (0) and (1) are the edges BD and AD respectively; the line (2′) passes through D and is parallel to BC; the line (3′) also passes through D and bisects the edge BC. The line (12′) passes through the middle point of AD and is parallel to BC; the line (13′) passes through the middle point of BC and is parallel to AD.

There is a binode at D and the edge is torsal. The tangent plane coincides with one of the biplanes; we have thus an ordinary biplane, and an oscular biplane (Plate 8).

31. Species VIII = 12 − 3C$_2$.

The equation of the surface is

$$y^3 + y(x+z+w) + 4axzw = 0,$$

where we set $$(m-1)^2 = 4am.$$

Consequently $$m_1 + m_2 = 2 + 4a,\ m_1 m_2 = 1.$$

Putting $a = \frac{1}{8}$, we obtain $m_1 = 2$, $m_2 = \frac{1}{2}$.

The equations of the twelve distinct lines upon the surface are as follows:

$$\begin{aligned}
(7):&\quad w=0,\ y=0,\\
(8):&\quad x=0,\ y=0,\\
(9):&\quad z=0,\ y=0,\\
(\overline{7}):&\quad y+z+x=0,\ w=0,\\
(\overline{8}):&\quad y+x+w=0,\ z=0,\\
(\overline{9}):&\quad y+z+w=0,\ x=0,\\
(1):&\quad 2y=2x=-z,\\
(2):&\quad 2y=-x=2z,\\
(3):&\quad 2y=-w=2x,\\
(4):&\quad 2y=2w=-x,\\
(5):&\quad 2y=2z=-w,\\
(6):&\quad 2y=-z=2w.
\end{aligned}$$

The lines (7), (8) and (9) are the edges CA, CD and AD, respectively. The lines $(\overline{7})$, $(\overline{8})$ and $(\overline{9})$ are the lines at infinity in the planes ABC, ABD, BCD, respectively. The lines (1) and (2) pass through the point D, and are parallel to the internal bisectors of the angles C and A, respectively, lying in the plane ABC. The lines (3) and (4) pass through the point C, and are parallel to the internal bisectors of the angles D and A, respectively, lying in the plane ABD.

The lines (5) and (6) pass through the point A, and are parallel to the internal bisectors of the angles D and C, respectively, lying in the plane BCD.

There are three conic nodes, at D, C, and A, respectively. The axes, each joining two nodes, are lines on the surface (Plate 9).

32. Species $\mathbf{IX = 12 - 2B_3}$.

The equation of the surface is

$$wxz + (a, b, c, d \between x, y)^3 = 0,$$

where we set

$$(a, b, c, d \between x, y)^3 = -d\,(f_1 x - y)\,(f_2 x - y)\,(f_3 x - y).$$

The equations of the seven distinct lines upon the surface are as follows:

$$\begin{aligned}
(0):&\quad x = 0,\ y = 0,\\
(1):&\quad f_1 x - y = 0,\ z = 0,\\
(2):&\quad f_2 x - y = 0,\ z = 0,\\
(3):&\quad f_3 x - y = 0,\ z = 0,\\
(4):&\quad f_1 x - y = 0,\ w = 0,\\
(5):&\quad f_2 x - y = 0,\ w = 0,\\
(6):&\quad f_3 x - y = 0,\ w = 0.
\end{aligned}$$

The following values were chosen for the constants:

$$f_1 = 1/3,\ f_2 = 1,\ f_3 = 3.$$

The line (0) is the edge CD; the line (2) bisects the interior angle D, in the plane ABD; the line (5) bisects the interior angle C, in the plane ABC. The data for the remaining lines are given in the following table:

	x	y	z	w
1 passes through	D in plane ABD			
meets AB	3	1		
3 passes through	D in plane ABD			
meets AB	1	3		
4 passes through	C in plane ABC			
meets AB	3	1		
6 passes through	C in plane ABC			
meets AB	1	3		

There are two binodes on the surface, at D and C, respectively. The axis, joining the two binodes, is a line on the surface (Plate 10).

33. Species $\mathbf{X = 12 - B_4 - C_2}$.

The equation of the surface is

$$wxz + (x + z)(y^2 - x^2) = 0.$$

The equations of the seven distinct lines upon the surface are as follows:

$$(0): \quad x = 0, \ y = 0,$$
$$(3): \quad x = 0, \ z = 0,$$
$$(1): \quad x - y = 0, \ z = 0,$$
$$(2): \quad x + y = 0, \ z = 0,$$
$$(1'): \quad x - y = 0, \ w = 0,$$
$$(2'): \quad x + y = 0, \ w = 0,$$
$$(12): \quad x + z = 0, \ w = 0.$$

The lines (0) and (3) are the edges CD and BD, respectively. The lines (1) and (2) lie in the plane ABD and bisect the angle D internally and externally, respectively. The lines (1′) and (2′) lie in the plane ABC and bisect the angle C internally and externally, respectively. The line (12) lies in the plane ABC and bisects externally the angle B.

There is a binode at D, a conic node at C; and the axis, joining the two nodes, is a line on the surface (Plate 8).

34. Species $\mathbf{XI = 12 - B_6}$.

The equation of the surface is

$$wxz + y^2z + x^3 - z^3 = 0.$$

The equations of the three distinct lines upon the surface are as follows:

$$(0): \quad x = 0, \ z = 0,$$
$$(1): \quad x = 0, \ y + z = 0,$$
$$(2): \quad x = 0, \ y - z = 0.$$

The line (0) is the edge BD; the lines (1) and (2) lie in the plane BCD, and bisect, externally and internally, the angle D.

There is a binode at D, and the tangent plane coinciding with one of the biplanes is oscular. We thus have an ordinary biplane, and an oscular biplane. The edge is a line on the surface (Plate 8).

35. Species $XII = 12 - U_6$.

The equation of the surface is

$$w(x+y+z)^2 + xyz = 0.$$

The equations of the six distinct lines upon the surface are as follows:

$$(1):\quad x = 0,\ y + z = 0,$$
$$(2):\quad y = 0,\ z + x = 0,$$
$$(3):\quad z = 0,\ x + y = 0,$$
$$(1'):\quad x = 0,\ w = 0,$$
$$(2'):\quad y = 0,\ w = 0,$$
$$(3'):\quad z = 0,\ w = 0.$$

The lines (1), (2), (3) pass through the point D, and are parallel to the lines BC, AC and AB, respectively. The lines (1′), (2′), (3′) are the edges BC, AC, AB, respectively.

There is a unode at D, i.e. the quadric cone has become a coincident plane-pair. This uniplane meets the cubic surface in three lines through the unode. Here these three lines or rays are distinct (Plate 11).

36. Species $XIII = 12 - B_3 - 2C_2$.

The equation of the surface is

$$wxz + y^2(x+y+z) = 0.$$

The equations of the eight distinct lines upon the surface are as follows:

$$(5):\quad x = 0,\ y = 0,$$
$$(6):\quad z = 0,\ y = 0,$$
$$(0):\quad y = 0,\ w = 0,$$
$$(1):\quad x = 0,\ y + z = 0,$$
$$(2):\quad z = 0,\ x + y = 0,$$
$$(3):\quad w = y = -z,$$
$$(4):\quad w = y = -x,$$
$$(012):\quad w = 0,\ x + y + z = 0.$$

The lines (5), (6) and (0) are the edges CD, AD and AC, respectively. The lines (1) and (2) pass through the point D, and are parallel to BC and AB, respectively. The line (3) passes through the point A, and meets the plane BCD at the point of intersection of the internal bisector of the angle C and the external bisectors of the angles B and D. The line (4) passes through the point C, and meets the plane ABD at the point of intersection of the internal bisector of the angle A and the external bisectors of the angles B and D. The line (012) lies at infinity in the plane ABC.

There is a binode at D, and two conic nodes at A and C, respectively. The axes, each joining the binode with a conic node, and the axis, joining the two conic nodes, are lines on the surface (Plate 11).

37. Species XIV $= 12 - B_5 - C_2$.

The equation of the surface is

$$wxz + y^2z + yx^2 = 0.$$

The equations of the four distinct lines upon the surface are as follows:

$$(0):\quad x = 0,\ y = 0,$$

$$(1):\quad x = 0,\ z = 0,$$

$$(2):\quad z = 0,\ y = 0,$$

$$(3):\quad w = 0,\ y = 0.$$

The lines (0), (1), (2), (3) are the edges CD, BD, AD, AC, respectively.

There is a binode at D, a conic node at C. The axis and the edge are lines on the surface (Plate 11).

38. Species XV $= 12 - U_7$.

The equation of the surface is

$$wx^2 + xz^2 + y^2z = 0.$$

The equations of the three distinct lines upon the surface are as follows:

$$(1):\quad x = 0,\ y = 0,$$

$$(2):\quad x = 0,\ z = 0,$$

$$(3):\quad z = 0,\ w = 0.$$

The lines (1), (2), (3) are the edges CD, BD, AB, respectively.

There is a unode at D, and two of the three rays in the uniplane BDC are coincident (Plate 11).

39. Species XVI = 12 − 4C$_2$.

The equation of the surface is

$$w(xy + xz + yz) + xyz = 0.$$

The equations of the nine distinct lines upon the surface are as follows:

$$\begin{aligned}
(12):&\quad z = 0,\ w = 0,\\
(13):&\quad y = 0,\ w = 0,\\
(14):&\quad y = 0,\ z = 0,\\
(23):&\quad x = 0,\ w = 0,\\
(24):&\quad x = 0,\ z = 0,\\
(34):&\quad x = 0,\ y = 0,\\
(12.34):&\quad x + y = 0,\ z + w = 0,\\
(13.24):&\quad x + z = 0,\ y + w = 0,\\
(14.23):&\quad x + w = 0,\ y + z = 0.
\end{aligned}$$

The lines (12), (13), (14), (23), (24), (34) are the edges AB, AC, AD, BC, BD, CD, respectively. The lines (12.34), (13.24), (14.23) meet the pairs of lines AB, CD; BD, AC; BC, AD, respectively, at infinity.

There are four conic nodes, at A, B, C and D; and the axes, each through two nodes, are lines on the surface (Plate 11).

40. Species XVII = 12 − 2B$_3$ − C$_2$.

The equation of the surface is

$$wxz + xy^2 + y^3 = 0.$$

The equations of the five distinct lines upon the surface are as follows:

$$\begin{aligned}
(0):&\quad x = 0,\ y = 0,\\
(1):&\quad y = 0,\ z = 0,\\
(2):&\quad y = 0,\ w = 0,\\
(3):&\quad z = 0,\ x + y = 0,\\
(4):&\quad w = 0,\ x + y = 0.
\end{aligned}$$

The lines (0), (1) and (2) are the edges CD, AD and AC, respectively. The lines (3) and (4) are parallel to the line AB, and pass through the points D and C, respectively.

There are two binodes, at D and C, respectively, and a conic node at A. The axis joining the two binodes, and the axes, each through the conic node and a binode, are lines on the surface (Plate 11).

41. Species $XVIII = 12 - B_4 - 2C_2$.

The equation of the surface is

$$wxz + y^2(x+z) = 0.$$

The equations of the five distinct lines upon the surface are as follows:

$$(1): \quad y=0, \ x=0,$$
$$(2): \quad y=0, \ z=0,$$
$$(0): \quad y=0, \ w=0,$$
$$(3): \quad x=0, \ z=0,$$
$$(4): \quad w=0, \ x+z=0.$$

The lines (1), (2), (0), (3) are the edges *CD*, *AD*, *AC*, *BD* respectively. The line (4) passes through *B* and is parallel to *AC*.

There is a binode at *D*, and two conic nodes at *A* and *C*, respectively. The axes, each through the binode and a conic node, the axis through the two conic nodes, and the edge of the binode are all lines on the surface (Plate 12).

42. Species $XIX = 12 - B_6 - C_2$.

The equation of the surface is

$$wxz + y^2z + x^3 = 0.$$

The equations of the two distinct lines upon the surface are as follows:

$$(1): \quad x=0, \ y=0,$$
$$(2): \quad x=0, \ z=0.$$

The lines (1) and (2) are the edges *CD* and *BD*, respectively.

There is a binode at *D*, a conic node at *C*. The axis, joining the binode and the conic node, and the edge of the binode are lines on the surface (Plate 12).

43. Species $XX = 12 - U_8$.

The equation of the surface is

$$x^2w + xz^2 + y^3 = 0.$$

The equation of the one distinct line upon the surface is

$$(1): \quad x=0, \ y=0.$$

The line (1) is the edge *CD*.

There is a unode at *D*, and the three rays in the uniplane are coincident. In this case, the line *CD* represents twenty-seven coincident straight lines (Plate 12).

44. Species XXI = 12 − 3B$_3$.

The equation of the surface is

$$wxz + y^3 = 0.$$

The equations of the three distinct lines upon the surface are as follows:

$$(1): \quad y = 0, \quad x = 0,$$

$$(2): \quad y = 0, \quad z = 0,$$

$$(3): \quad y = 0, \quad w = 0.$$

The lines (1), (2), (3) are the edges CD, AD, AC, respectively.

There are three binodes, at D, C and A, respectively. The axes, each joining two binodes, are lines on the surface (Plate 12).

CHAPTER VII

ON SOME CONFIGURATIONS ASSOCIATED WITH THE CONFIGURATIONS OF THE LINES UPON THE CUBIC SURFACE

45. Concerning the Brianchon Configuration.

Cayley* has considered the question of deriving the Pascalian configuration, by projection, from a pair of triheders. Denote the three planes of one triheder by a_1, a_2, a_3; of the other by b_1, b_2, b_3. Considering the nine lines $a_i b_j$ $(i, j = 1, 2, 3)$ and taking them in a particular way in six sets of three each, we may pass hyperboloids through each set of three lines. These hyperboloids intersect in four points O_1, O_2, O_3, O_4; and if we project the solid figure of the two triheders from any one of these four points upon an arbitrary plane, the resulting figure is the Pascalian configuration. The polar planes of any one of the points O with respect to each one of the triheders are identical.

There is, of course, by the principle of duality, a corresponding theorem for two point-triads in space. The proof of this theorem may be effected directly in a very simple manner, by use of the equations already derived in a former article. Choosing the equations in the manner given below, we gain the advantage of bringing the critical plane to coincide with the plane of one of the faces of the fundamental tetrahedron, thereby furnishing a basis of facile procedure. This will appear in the sequel.

Four point co-ordinates are used, an equation of the form

$$xu_1 + yu_2 + zu_3 + wu_4 = 0$$

being the equation of a point, the co-ordinates being the variables u_1, u_2, u_3, u_4, which represent the perpendiculars, from the four points A,

* *Coll. Math. Papers*, Vol. VI. pp. 129–134; *Quart. Journ.* Vol. IX. (1868), pp. 348–353.

B, C, D of the fundamental tetrahedron $ABCD$, upon any plane passing through the point in question.

Consider two point-triads in space, the one triad consisting of the points designated L, M, N; the other consisting of the points designated P, Q, R. The equations of the points are chosen as below; the results are precisely the same, had the equations of the points been chosen in all their generality (Cayley's paper, l.c.).

$$\begin{cases} L: & u_4 = 0, \\ M: & mu_1 + lu_2 + lmnu_3 + (mn - l)(nl - m)(lmn - 1)u_4 = 0, \\ N: & nlu_1 + mnu_2 + u_3 - (mn - l)(nl - m)(lmn - 1)u_4 = 0, \end{cases}$$

and

$$\begin{cases} P: & u_1 - (mn - l)(lmn - 1)u_4 = 0, \\ Q: & u_2 - (nl - m)(lmn - 1)u_4 = 0, \\ R: & u_3 + (mn - l)(nl - m) \ u_4 = 0. \end{cases}$$

Let v_1, v_2, v_3, v_4 denote constant values of u_1, u_2, u_3, u_4, respectively. Also set

$$\left.\frac{\partial F}{\partial u_1}\right]_{u_1 = v_1} = \frac{\partial F}{\partial v_1},$$

and similarly in other cases.

The initial problem is to find a plane such that its pole with respect to the system of points, written in the symbolic form

$$LMN = 0 \quad \dots\dots\dots\dots\dots\dots\dots\dots\dots(1),$$

is identical with its pole with respect to the second system of three points, written

$$PQR = 0 \quad \dots\dots\dots\dots\dots\dots\dots\dots\dots(2),$$

The pole $$\frac{\partial F}{\partial v_1}u_1 + \frac{\partial F}{\partial v_2}u_2 + \frac{\partial F}{\partial v_3}u_3 + \frac{\partial F}{\partial v_4}u_4 = 0$$

of the plane (v_1, v_2, v_3, v_4) with respect to the system (1) given by the equation

$$F(u_1, u_2, u_3, u_4) = 0$$

has for its equation

$$\begin{aligned} &[2lmn\, v_1v_4 + n(l^2 + m^2)v_2v_4 + m(l^2n^2 + 1)v_3v_4 + \lambda\mu^2\nu \,.\, v_4^2]\,u_1 \\ &+ [2lmn\, v_2v_4 + n(l^2 + m^2)v_1v_4 + l(m^2 + n^2)v_3v_4 + \lambda^2\mu\nu \,.\, v_4^2]\,u_2 \\ &+ [2lmn\, v_3v_4 + m(l^2n^2 + 1)v_1v_4 + l(m^2 + n^2)v_2v_4 - \lambda\mu\nu^2 \,.\, v_4^2]\,u_3 \\ &+ [lmn(v_1^2 + v_2^2 + v_3^2) - 3\lambda^2\mu^2\nu^2 \,.\, v_4^2 + n(l^2 + m^2)v_1v_2 + m(l^2m^2 + 1)v_1v_3 \\ &+ l(m^2 + n^2)v_2v_3 + 2\lambda\mu^2\nu \,.\, v_1v_4 + 2\lambda^2\mu\nu \,.\, v_2v_4 - 2\lambda\mu\nu^2 \,.\, v_3v_4]\,u_4 = 0 \dots(3), \end{aligned}$$

where $\lambda, \mu, \nu \equiv mn - l, \ nl - m, \ lmn - 1$, respectively.

Also, the pole

$$\frac{\partial \Phi}{\partial v_1} u_1 + \frac{\partial \Phi}{\partial v_2} u_2 + \frac{\partial \Phi}{\partial v_3} u_3 + \frac{\partial \Phi}{\partial v_4} u_4 = 0$$

of the plane (v_1, v_2, v_3, v_4) with respect to the system (2) given by the equation

$$\Phi(u_1, u_2, u_3, u_4) = 0$$

has for its equation

$$\begin{aligned}
&[v_2v_3 - \mu\nu \,.\, v_3v_4 + \lambda\mu \,.\, v_2v_4 - \lambda\mu^2\nu \,.\, v_4^2]\, u_1 \\
&+ [v_1v_3 - \lambda\nu \,.\, v_3v_4 + \lambda\mu \,.\, v_1v_4 - \lambda^2\mu\nu \,.\, v_4^2]\, u_2 \\
&+ [v_1v_2 - \lambda\nu \,.\, v_2v_4 - \mu\nu \,.\, v_1v_4 + \lambda\mu\nu^2 \,.\, v_4^2]\, u_3 \\
&+ [-\lambda\nu \,.\, v_2v_3 - \mu\nu \,.\, v_1v_3 + 2\lambda\mu\nu^2 \,.\, v_3v_4 + \lambda\mu \,.\, v_1v_2 \\
&- 2\lambda^2\mu\nu \,.\, v_2v_4 - 2\lambda\mu^2\nu \,.\, v_1v_4 + 3\lambda^2\mu^2\nu^2 \,.\, v_4^2]\, u_4 = 0 \quad \text{.........(4)},
\end{aligned}$$

where $\lambda, \mu, \nu \equiv mn - l,\ nl - m,\ lmn - 1$, respectively,

as before.

Now, it is evident by inspection that equations (3) and (4) are identical (aside from sign) if

$$v_1 = v_2 = v_3 = 0.$$

Accordingly the plane of the face ABC of the fundamental tetrahedron $ABCD$ is such that its pole with respect to the point-triad (1) is coincident with its pole with respect to the point-triad (2).

Connect up next the six points L, M, N, P, Q, R by lines and planes in every possible way. Suppose the plane of ABC to be intersected by the line LM in the point LM, and by the plane LMN in the line LMN; and so in other cases. We obtain in this fashion a configuration in the plane of ABC, consisting of the fifteen ($\equiv {}_2C_6$) points $LM, LN, \ldots QR$, and of the twenty ($\equiv {}_3C_6$) lines $LMN, LMP, \ldots PQR$; and which is such that through each of the points there pass four of the lines, and on each of the lines lie three of the points. Thus the lines

$$\left.\begin{matrix} LMN \\ LMP \\ LMQ \\ LMR \end{matrix}\right\} \text{pass through the point } LM;$$

and the points

$$\left.\begin{matrix} LM \\ MN \\ NL \end{matrix}\right\} \text{lie on the line } LMN;$$

and so in other cases.

It will next be shown that six lines, denoted by 1, 2, 3, 4, 5, 6, may be drawn in the plane ABC, conditioned as follows:

$$(A): \begin{cases} \text{line (1) passes through the points } LP,\ MQ,\ NR, \\ \quad ,, \ (2) \quad ,, \quad ,, \quad ,, \quad ,, \quad LQ,\ MR,\ NP, \\ \quad ,, \ (3) \quad ,, \quad ,, \quad ,, \quad ,, \quad LR,\ MP,\ NQ, \\ \quad ,, \ (4) \quad ,, \quad ,, \quad ,, \quad ,, \quad LP,\ MR,\ NQ, \\ \quad ,, \ (5) \quad ,, \quad ,, \quad ,, \quad ,, \quad LQ,\ MP,\ NR, \\ \quad ,, \ (6) \quad ,, \quad ,, \quad ,, \quad ,, \quad LR,\ MQ,\ NP. \end{cases}$$

For this purpose, represent any line in the plane of ABC as the join of two points, whose equations are

$$\left.\begin{aligned} \lambda_1 u_1 + \mu_1 u_2 + \nu_1 u_3 = 0 \\ \lambda_2 u_1 + \mu_2 u_2 + \nu_2 u_3 = 0 \end{aligned}\right\}.$$

If, for example, this line meets the line LP, the join of the two points, whose equations are

$$\begin{aligned} L:&\quad u_4 = 0, \\ P:&\quad u_1 - (mn - l)(lmn - 1)\, u_4 = 0, \end{aligned}$$

we have the equation of condition

$$\begin{vmatrix} \lambda_1, & \mu_1, & \nu_1, & 0 \\ \lambda_2, & \mu_2, & \nu_2, & 0 \\ 0, & 0, & 0, & 1 \\ 1, & 0, & 0, & -(mn-l)(lmn-1) \end{vmatrix} = 0,$$

or

$$\mu_1 : \mu_2 = \nu_1 : \nu_2 ;$$

and hence the line in question may be written

$$\left.\begin{aligned} u_1 = 0 \\ \mu_1 u_2 + \nu_1 u_3 = 0 \end{aligned}\right\}.$$

If further, this line meets the line MQ, we have the equation of condition

$$\begin{vmatrix} 1, & 0, & 0, & 0 \\ 0, & \mu_1, & \nu_1, & 0 \\ m, & l, & lmn, & (mn-l)(nl-m)(lmn-1) \\ 0, & 1, & 0, & -(nl-m)(lmn-1) \end{vmatrix} = 0,$$

or

$$\mu_1 : \nu_1 = 1 : l,$$

and hence the required line has for its equations

$$1: \quad \begin{cases} u_1 = 0, \\ u_2 + l u_3 = 0. \end{cases}$$

If now we write the equation of the point N in the form

$$N: \quad (nlu_1 + mnu_2 + lmnu_3) - (lmn - 1)\{u_3 + (mn - l)(nl - m)\, u_4\} = 0,$$

and note the equation of the point R:

$$R: \quad u_3 + (mn - l)(nl - m)\, u_4 = 0,$$

it is evident that the equations of the line NR may be written

$$NR: \quad \begin{Bmatrix} nlu_1 + mnu_2 + lmnu_3 = 0 \\ u_3 + (mn - l)(nl - m)\, u_4 = 0 \end{Bmatrix}.$$

That the line (1) meets the line NR is now obvious by inspection.

Determining, in similar fashion, the equations of the five remaining lines, and re-writing the equations of line (1), we obtain

$$1: \quad \begin{Bmatrix} u_1 = 0 \\ u_2 + lu_3 = 0 \end{Bmatrix},$$

$$2: \quad \begin{Bmatrix} u_2 = 0 \\ mu_1 + u_3 = 0 \end{Bmatrix},$$

$$3: \quad \begin{Bmatrix} u_3 = 0 \\ nu_1 + u_2 = 0 \end{Bmatrix},$$

$$4: \quad \begin{Bmatrix} u_1 = 0 \\ lu_2 + u_3 = 0 \end{Bmatrix},$$

$$5: \quad \begin{Bmatrix} u_2 = 0 \\ u_1 + mu_3 = 0 \end{Bmatrix},$$

$$6: \quad \begin{Bmatrix} u_3 = 0 \\ u_1 + nu_2 = 0 \end{Bmatrix}.$$

Now these six lines 1, 2, 3, 4, 5, 6 touch the conic given by the equation

$$lmn\,(u_1^2 + u_2^2 + u_3^2) + mn\,(l^2 + 1)\,u_2u_3 + nl\,(m^2 + 1)\,u_3u_1 + lm\,(n^2 + 1)\,u_1u_2 = 0.$$

This is most easily shown by putting u_1, u_2, u_3 in turn equal to zero in the equation last written. We obtain, respectively,

$$mn\,(u_2 + lu_3)\,(lu_2 + u_3) = 0,$$
$$nl\,(mu_1 + u_3)\,(u_1 + mu_3) = 0,$$
$$lm\,(nu_1 + u_2)\,(u_1 + nu_2) = 0.$$

Moreover, it is clear from an inspection of the scheme (A) above, that the points LP, LQ, LR; MP, MQ, MR; NP, NQ, NR are the points 14, 25, 36; 35, 16, 24; 26, 34, 15, respectively, where 14, for example, denotes the meet of the lines 1 and 4; and so in other cases.

Conversely, starting from the six lines 1, 2, 3, 4, 5, 6 touching the conic, and denoting the points 14, 25, 36; 35, 16, 24; 26, 34, 15 (which are, indeed, the vertices, and meets of opposite sides, of the hexad 162435) in the manner described above, then it is possible to complete the figure of the fifteen points LM, LN,...QR, and of the twenty lines LMN, LMP,...PQR, such that through each point pass four lines, and on each line lie three points, as detailed in the foregoing.

Of the fifteen points, nine, viz. the points LP, LQ, LR; MP, MQ, MR; NP, NQ, NR are, as appeared above, points on two of the six lines 1, 2, 3, 4, 5, 6; the remaining points are MN, NL, LM; QR, RP, PQ. These are *Brianchon Points*:

MN of the hexad 162435,
NL ,, ,, ,, 152634,
LM ,, ,, ,, 142536,
QR ,, ,, ,, 152436,
RP ,, ,, ,, 142635,
PQ ,, ,, ,, 162534,

for the point MN is the meet of the lines MNP, MNQ, $MNR \equiv MP$, NP; MQ, NQ; MR, $NR \equiv 35$, 26; 16, 34; 24, 15; that is, MN is the Brianchon point of the hexad 162435; and similar reasoning verifies the above statements for the remainder of the six points.

To summarize, we have two sets of three hexads, such that the Brianchon points of each set lie *in linea*; and the two lines so obtained, together with the eighteen lines through the six Brianchon points, form a system of twenty lines passing by fours through fifteen points (Fig. 2).

46. Derivation of the Pascalian Configuration by the Projection of the Straight Lines lying upon the Cubic Surface of the Second Species.

For the cubic surface with one conical point (§ 25), the tangent planes of the type [12 . 34 . 56] are fifteen in number, viz.

[12 . 34 . 56], [14 . 23 . 56], [15 . 23 . 46],
[12 . 36 . 45], [13 . 24 . 56], [15 . 24 . 36],
[12 . 35 . 46], [16 . 23 . 45], [13 . 25 . 46],
[16 . 25 . 34], [13 . 26 . 45], [14 . 26 . 35],
[15 . 26 . 34], [16 . 24 . 35], [14 . 25 . 36].

In terms of these fifteen planes, the equation of the surface may be written in the form

$$UVW + kXYZ = 0$$

in ten different ways (§ 13, third table of trihedral pairs).

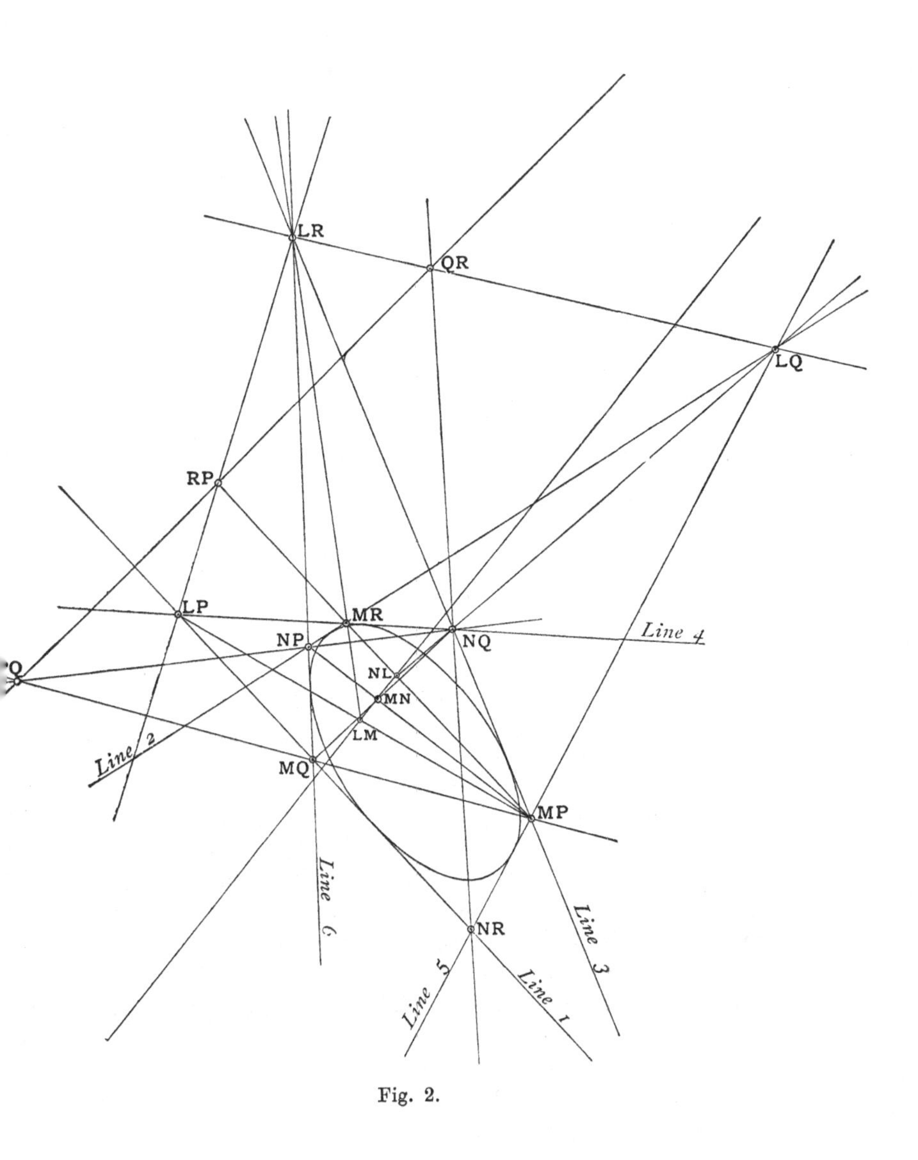
LR
QR
LQ
RP
LP
MR
NQ
Line 4
NP
NL
MN
LM
Line 2
MQ
MP
Line 6
NR
Line 3
Line 5
Line 1

Fig. 2.

Consider one of these forms, viz.

$$[12.34.56]\times[15.23.46]\times[14.26.35] + k[12.35.46]\times[15.26.34]\times[14.23.56]=0,$$

written in the symbolic notation. This conjugate trihedral pair cuts out from the cubic surface the nine lines

12, 14, 15; 23, 34, 35; 26, 46, 56.

Now let us consider the six lines

1, 2, 3, 4, 5, 6

upon the cubic surface, which all pass through one point O, the vertex of the quadric cone upon which they lie. It is clear by inspection that

(A): the line 1 meets the lines 12, 15, 14,
,, ,, 3 ,, ,, ,, 34, 23, 35,
,, ,, 6 ,, ,, ,, 56, 46, 26,
,, ,, 2 ,, ,, ,, 12, 23, 26,
,, ,, 4 ,, ,, ,, 34, 46, 14,
,, ,, 5 ,, ,, ,, 56, 15, 35,

For the sake of brevity, I shall designate the six planes as follows:

$$[12.34.56]\equiv a,\quad [12.35.46]\equiv f,$$
$$[15.23.46]\equiv b,\quad [15.26.34]\equiv g,$$
$$[14.26.35]\equiv c,\quad [14.23.56]\equiv h.$$

Consider now the six planes a, b, c, f, g, h, and taking O as the point of projection, and an arbitrary plane of projection, the line of intersection of the planes a and b will be projected into a line ab, and the point of intersection of the planes a, b, c into a point abc; and so in other cases. We have thus a plane figure consisting of the fifteen lines $ab, ac, \ldots gh$, and of the twenty points $abc, abf, \ldots fgh$; and which is such that on each of the lines there lie four of the points, and through each of the points there pass three of the lines, viz. the points abc, abf, abg, abh lie on the line ab; and the lines bc, ca, ab meet in the point abc; and so in other cases.

Moreover, from the above scheme, we see that the projections of the lines af, bg, ch meet in a point, and the like for each of the six triads of lines; that is, in the plane figure, we have six points 1, 3, 6, 2, 4, 5—each of them the intersection of three lines as shown in the following scheme:

$$1 = af \,.\, bg \,.\, ch,$$
$$3 = ag \,.\, bh \,.\, cf,$$
$$6 = ah \,.\, bf \,.\, cg,$$
$$2 = af \,.\, bh \,.\, cg,$$
$$4 = ag \,.\, bf \,.\, ch,$$
$$5 = ah \,.\, bg \,.\, cf,$$

and these six points lie in a conic (the intersection of the quadric cone by the plane of projection). It is clear that the lines *af*, *ag*, *ah* ; *bf*, *bg*, *bh* ; *cf*, *cg*, *ch* are the lines 12, 34, 56 ; 46, 15, 23 ; 35, 26, 14, respectively.

Of the fifteen lines, nine, viz. the lines *af*, *ag*, *ah* ; *bf*, *bg*, *bh* ; *cf*, *cg*, *ch* are, as has been seen, lines through two of the six points 1, 3, 6, 2, 4, 5 ; the remaining lines are *bc*, *ca*, *ab* ; *gh*, *hf*, *fg*. These are Pascalian lines :

bc	for	the	hexagon	153264,
ca	,,	,,	,,	143562,
ab	,,	,,	,,	123465,
gh	,,	,,	,,	143265,
hf	,,	,,	,,	123564,
fg	,,	,,	,,	153462.

This appears as follows :

line *bc* contains points *bcf*, *bcg*, *bch*

$$= bf \,.\, cf,\ bg \,.\, cg,\ bh \,.\, ch$$
$$= 46 \,.\, 35,\ 15 \,.\, 26,\ 23 \,.\, 14\ ;$$

that is, *bc* is the Pascalian line of the hexagon 153264 ; and similarly in other cases.

The twenty points *abc*, *abf*, ... *fgh* are as follows, viz. omitting the two points *abc*, *fgh*, the remaining eighteen points are the Pascalian points (the intersections of pairs of lines each through two of the points 1, 2, 3, 4, 5, 6) which lie on the Pascalian lines *bc*, *ca*, *ab* ; *gh*, *hf*, *fg* respectively ; the point *abc* is the intersection of the Pascalian lines *bc*, *ca*, *ab* ; and the point *fgh* is the intersection of the Pascalian lines *gh*, *hf*, *fg*—the points in question being two of the points S (Steiner's twenty points, each the intersection of three Pascalian lines).

In this process, we have projected only a single one of the ten possible trihedral pairs. This projection of a single trihedral pair gave six Pascalian lines and two Steiner points. Remembering that the six points and the conic upon which they lie are fixed, in the process of projection, we reach the conclusion that the projection of all ten trihedral pairs upon the same plane give 60 Pascalian lines and 20 Steiner points, as should be the case.

It would not be difficult at this point to develop the theory so as to put in evidence the 60 Kirkman points, the twenty Salmon-Cayley lines, the fifteen Salmon points, and the fifteen Steiner lines for the plane configuration, derived by projection, from the point O, of corresponding elements in the spatial configuration. This was done by Cremona*, to whom the theorem is due, in his original paper on the subject. The subject has also been considered by Richmond†, who succeeded in giving a perfectly symmetrical form to the equations of the lines on this type of cubic surface, after Segre's method (cf. historical summary).

47. On the Graphic Representation of the Projection of a Pair of Triheders into the Pascalian Configuration.

The problem with which I have concerned myself here is: Can we represent to the eye, graphically or by means of a model, the figure arising from the projection of a pair of triheders into the Pascalian configuration?

I select the two triheders as follows:

$$\left.\begin{array}{l}(12\,.\,34\,.\,56)\,(15\,.\,23\,.\,46)\,(14\,.\,26\,.\,35)=0\\(12\,.\,35\,.\,46)\,(15\,.\,26\,.\,34)\,(14\,.\,23\,.\,56)=0\end{array}\right\}\,\ddagger.$$

These mutually intersect in the nine lines

$$12,\ 14,\ 15\ ;\ 23,\ 34,\ 35\ ;\ 26,\ 46,\ 56.$$

Using the numerical values given in § 25, these planes have the equations

$$\begin{aligned}a:&\quad w=0,\\b:&\quad 256x-384y-96z+459w=0,\\c:&\quad 96x-64y-256z+153w=0,\end{aligned}$$

and

$$\begin{aligned}f:&\quad 32x+27w=0,\\g:&\quad 64y-51w=0,\\h:&\quad 32z-17w=0.\end{aligned}$$

* "Teoremi Stereometrici dai quali si deducono le Proprieta dell' Esagrammo di Pascal," *Reale Accademia dei Lincei*, Anno LXXIV. (1876–77).

† *Quart. Journ.* Vol. XXIII. (1889), pp. 170–179.

‡ Cf. § 13.

Writing down the polar planes of these two triads *abc*, *fgh*, the condition for their coincidence requires their common pole O to assume, as one of its four possible positions, the position of the vertex D of the fundamental tetrahedron $ABCD$.

The lines through the point O, and conditioned by the scheme (A) of § 46, are the lines 1, 2, 3, 4, 5, 6 passing through the conical node of that article, which coincides with the vertex D of the fundamental tetrahedron $ABCD$.

The nine lines 12, 14, 15; 23, 34, 35; 26, 46, 56, and also the six lines 1, 2, 3, 4, 5, 6, are laid down on the diagram, precisely as was done for the Species II (§ 25) of the cubic surface. The point D is the point of projection, while the plane of projection, which is *within our choice*, may be taken to coincide with the plane of ABC, since this plane contains the three lines 12, 34, 56, but does not contain any one of the lines 1, 2, 3, 4, 5, 6.

The projection of the line 46, for example, was found by joining the meets of the lines 4 and 46 with the plane of projection (since the line 46 meets the line 4); and similarly for other cases. Only three of the Pascalian lines, viz. *bc*, *ca*, *ab* are shown in the figure, in order to avoid too great complexity. The projection of line *bc*, for example, was found as follows: lines *bh* and *ch* (i.e. lines 23 and 14), lying in planes *b* and *c* respectively, intersect in a point P say, on the line *bc*; and similarly the lines *bg* and *cg* (i.e. lines 15 and 26) intersect in a point P_1 say, on the line *bc*. The projections of the points P and P_1 are the meets of the projections of the pairs of lines *bh*, *ch*; *bg*, *cg* respectively. The projection of the line *bc*, then, is the join of the projections of the points P and P_1.

The Steiner point shown in the figure, the common meet of the three Pascalian lines *bc*, *ca*, *ab*, is one of the two Steiner points yielded by the projection; the other is not shown as explained above.

One other detail was the construction of the conic section (in the diagram an hyperbola) given by the intersection of the quadric cone (containing the lines 1, 2, 3, 4, 5, 6) with the plane of projection. The six points I, II, III, IV, V, VI (the meets of the lines 1, 2, 3, 4, 5, 6, respectively, with the plane of projection) lie in this conic section. Hence it was constructed, projectively, by means of Pascal's Theorem (Plate 13).

This graphic (or modelled) representation of a remarkable configuration and its projection relates not only to Cayley's configuration, but also to the Species II of the cubic surface. Indeed, it may be interpreted as a planographic representation of the projection of the lines upon a cubic surface having only one conical node, from the nodal point, upon an arbitrary plane into the projection of the "mystic hexagram."

It is perhaps worthy of remark that the six lines 1, 2, 3, 4, 5, 6 lie upon the quadric cone having D for its vertex, and for its base the conic section (an hyperbola) having for its equations

$$12(x^2+y^2+z^2)-5(6yz-8zx+5xy)=0, \quad w=0.$$

The intersections of this cone by the planes $x=0$, $y=0$, $z=0$, respectively, have for their equations:

$$x=0, \quad (2y-z)(y-2z)=0;$$
$$y=0, \quad (x+3z)(3x+z)=0;$$
$$z=0, \quad (3x+4y)(4x+3y)=0.$$

48. A Deduction from Cayley's Theorem on the Pascalian Configuration.

The theorem of Cayley* mentioned in § 45, together with the argument in § 46, leads to the well-known conclusion:

Given any two triads of planes a, b, c; f, g, h; then it is possible to find four points O_1, O_2, O_3, O_4 such that the polar plane of any one of these points with respect to one triheder is identical with its polar plane with respect to the other triheder. Considering any one of the points, say O_1, then it is possible to draw six lines through O_1, whose positions are defined as follows:

$$(A): \begin{cases} \text{line 1 meets the lines } af, bg, ch, \\ \;\;\text{,,}\;\; 2 \;\;\text{,,}\;\; \text{,,}\;\; \text{,,}\;\; ag, bh, cf, \\ \;\;\text{,,}\;\; 3 \;\;\text{,,}\;\; \text{,,}\;\; \text{,,}\;\; ah, bf, cg, \\ \;\;\text{,,}\;\; 4 \;\;\text{,,}\;\; \text{,,}\;\; \text{,,}\;\; af, bh, cg, \\ \;\;\text{,,}\;\; 5 \;\;\text{,,}\;\; \text{,,}\;\; \text{,,}\;\; ag, bf, ch, \\ \;\;\text{,,}\;\; 6 \;\;\text{,,}\;\; \text{,,}\;\; \text{,,}\;\; ah, bg, cf. \end{cases}$$

Then these six lines together with the nine lines af, ag, ah; bf, bg, bh; cf, cg, ch determine a cubic surface upon which they lie, for which the point O_1 is the only conical point.

This conclusion may be more generally phrased as follows:

Through the nine lines of mutual intersection of two triheders can be drawn four cubic surfaces, each possessing only one conical point

* *Coll. Math. Papers*, Vol. VI. pp. 129–134.

and having twenty-one distinct lines lying wholly upon the surface. The four points O_1, O_2, O_3, O_4 such that the plane of any one of them with respect to one triheder is identical with its polar plane with respect to the other triheder, are the conical points of the four cubic surfaces. Through each one of these points pass six lines, conditioned by the scheme (A) above, which lie not only upon a quadric cone but also upon the cubic surface through the nine lines above mentioned and for which that point is the only conical point.

A BIBLIOGRAPHY

OF

BOOKS AND PAPERS REFERRING TO THE SUBJECT OF THE PRESENT MEMOIR.

1849 "On the triple tangent planes of surfaces of the third order." By A. Cayley. *Camb. and Dublin Math. Journal*, Vol. IV. pp. 118–132.

"On the triple tangent planes to a surface of the third order." By G. Salmon. *Camb. and Dublin Math. Journal*, Vol. IV. pp. 252–260.

Compare also Salmon's "Geometry of Three Dimensions."

1855 "Intorno ad alcune proprieta delle superficie del terzo ordine." By F. Brioschi. *Annali di scienze mat. e fis.*, Roma.

1856–7 "The twenty-seven real straight lines on the cubic surface." By J. Steiner. *Monatsberichte der K. Preuss. Akademie der Wissenschaften*, Berlin, pp. 50....

Compare also Steiner's paper in *Crelle's Journal*, t. LIII. pp. 133–141.

1858 "An attempt to determine the twenty-seven lines upon a surface of the third order, and to divide such surfaces into species, in reference to the reality of the lines upon the surface." By L. Schläfli. *Quart. Journ.* Vol. II. pp. 55–65 and 110–120.

1859 "The twenty-seven straight lines on the cubic surface." By E. de Jonquières. *Nouvelles Annales de Mathématique*, Paris, Vol. XVIII. pp. 129....

1861 "Note sur l'involution de six lignes dans l'espace." By J. J. Sylvester. *Comptes Rendus*, Vol. LII. pp. 815–817.

"Note sur les 27 droites d'une surface du troisième degré." By J. J. Sylvester. *Comptes Rendus*, Vol. LII. pp. 977–980.

1862 "Disquisitiones de superficiebus tertii ordinis." By F. August. Dissert. inaug. Berolini.

1863 "On the distribution of surfaces of the third order into species, in reference to the presence or absence of singular points, and the reality of their lines." By L. Schläfli. *Philos. Trans.* Vol. CLIII. pp. 193–241.

"Nachweis der 27 Geraden auf der allgemeinen Oberfläche dritter Ordnung." By H. E. Schröter. *Crelle's Journal*, Vol. LXII. pp. 265....

1867 "Synthetische Untersuchungen über Flächen dritter Ordnung." By R. Sturm. B. G. Teubner, Leipzig.

1868 "Mémoire de géométrie pure sur les surfaces du troisième ordre." By L. Cremona. *Crelle's Journal*, Vol. LXVIII. pp. 1–133.

"A 'Smith's Prize' Paper; Solutions." By A. Cayley. *Coll. Math. Papers*, Vol. VIII. pp. 414–435.

Compare also *Coll. Math. Papers*, Vol. VI. pp. 129–134; *Quart. Journ.* Vol. IX. (1868), pp. 348–353.

1869 "Ueber die Doppeltangenten einer ebenen Curve vierten Grades." By C. F. Geiser. *Math. Ann.* Bd. I. pp. 129–138.

"On the six co-ordinates of a line." By A. Cayley. *Trans. Camb. Philos. Soc.* Vol. XI. Part II. pp. 290–323.

"A Memoir on Cubic Surfaces." By A. Cayley. *Philos. Trans. Royal Soc. of London*, Vol. CLIX. pp. 231–326.

"The equation of the twenty-seven lines upon the cubic surface." By C. Jordan. *Liouville's Journ. Math.* Vol. XIV. pp. 147....

Compare also *Comptes Rendus*, Vol. LXVIII. pp. 865....

1870 "Sulle ventisette rette di una superficie del terzo ordine." By L. Cremona. *Rendiconti dell' Istituto Lombardo*, Ser. 2, Vol. III. pp. 209....

"On the double-sixers of a cubic surface." By A. Cayley. *Coll. Math. Papers*, Vol. VII. pp. 316–329; *Quart. Journ.* Vol. X. pp. 58–71.

"Sur une nouvelle combinaison des 27 droites d'une surface du troisième ordre." By C. Jordan. *Comptes Rendus*, Vol. LXX. pp. 326–328.

1871 "Quand è che dalla superficie generale di terz' ordine si stacca una parte che non sia realmente segata da ogni piano reale?" By L. Schläfli. *Annali di Mat.* Vol. V. (II.), pp. 289–295.

1873 "The triple tangent planes to the cubic surface." By W. Spottiswoode. *Comptes Rendus*, Vol. LXXVII. pp. 1181....

"Ueber Flächen dritter Ordnung." By F. Klein. *Math. Ann.* Bd. VI. pp. 551–581.

"Extension to cubic surfaces of Pascal's and Brianchon's theorems." By F. Folie, *Mémoires de la Société Royale des Sciences, de l'Agriculture, et des Arts*, Liège, Vol. III. pp. 663....

"On Dr Wiener's Model of a Cubic Surface with 27 Real Lines; and on the Construction of a Double-Sixer." By A. Cayley. *Trans. Camb. Philos. Soc.* Vol. XII. Part I. pp. 366–383.

1874 "Sur les différentes formes des courbes planes du quatrième ordre." By H. G. Zeuthen. *Math. Ann.* Vol. VI. pp. 410–432.

"The twenty-seven lines on the cubic surface." By F. G. Affolter. *Archiv der Math. und Phys.*; Grunert, Greifswald, Leipzig, Vol. LVI. pp. 113....

1875 "Études des propriétés de situation des surfaces cubiques." By H. G. Zeuthen. *Math. Ann.* Bd. VIII. pp. 1–30.

1876 "Geometrical Instruments and Models." By H. J. S. Smith. *South Kensington Museum Handbook to the Special Loan Collection of Scientific Apparatus*, pp. 34–54.

"A property of the triple tangent planes." By F. Brioschi. *Atti della Reale Accademia dei Lincei*, Roma, Vol. III. (Pte. II.), pp. 257....

"Teoremi Stereometrici dai quali si deducono le Proprieta dell' Esagrammo di Pascal." By L. Cremona. *Reale Accademia dei Lincei*, Anno LXXIV., Roma.

1879 "Zur Classification der Flächen dritter Ordnung." By C. Rodenberg. *Math. Ann.* Bd. XIV. pp. 46–110. Signed "Im December 1877."

"On Double-Sixers." By A. Cayley. *Trans. Camb. Philos. Soc.* Vol. XII. pp. 366....

1880 "Delineation of the twenty-seven lines upon the cubic surface." By J. Carou. *Bull. de la Société Math. de France*, Paris, Vol. VIII. pp. 73....

1881 "Ueber die durch collineare Grundgebilde erzeugten Curven und Flächen." By F. Schur. *Math. Ann.* Bd. XVIII. pp. 1–32.

1882 "On the 27 lines, the 45 triple tangent planes, and the 36 double-sixers of a cubic surface, with a hint for the construction of models which give the position of the lines when they are all real." By P. Frost. *Quart. Journ.* Vol. XVIII. pp. 89–96.

1883 "The twenty-seven lines upon the cubic surface, and the parabolic curve." By G. Bauer. *Münch. Akademie Sitzungberichte*, Vol. XIII. pp. 320....

"Contribuzione alla teoria delle 27 rette e dei 45 piani tritangenti di una superficie di 3° ordine." By E. Bertini. *Annali di Matematica*, Vol. XII. (Part II.), pp. 301–346.

1884 "Ueber die 27 Geraden der cubischen Fläche." By R. Sturm. *Math. Ann.* Vol. XXIII. pp. 289–310.

"Polyhedral configurations of the triple tangent planes." By E. Bertini. *Reale Istituto Lombardo, Rendiconti*, Milan, Vol. XVII. pp. 478..., 712....

1886 "Extension to cubic surfaces of Pascal's and Brianchon's theorems." By A. Petot. *Comptes Rendus*, Vol. CII. pp. 737....

1889 "A symmetrical system of equations of the lines on a cubic surface which has a conical point." By H. W. Richmond. *Quart. Journ.* Vol. XXIII. pp. 170–179.

1891 "Case of hexad of lines on the cubic surface." By G. Kohn. *Monatshefte für Math. und Phys.*, Wien, Vol. II. pp. 293....

"Proof of Cayley's theorem on the triple tangent planes to the cubic surface." By G. Kohn. *Monatshefte für Math. und Phys.*, Wien, Vol. II. pp. 343....

1893 "Construction of models of cubic surfaces by Rodenberg." By D. J. Korteweg. *Nieuw Archief voor Wiskunde*, Amsterdam, Vol. XX. pp. 63....

Compare also *Fortschritte der Math.* (1893–4), pp. 83....

"Representation on the plane of the twenty-seven lines upon a cubic surface." By P. H. Schoute. *Amsterdam Akademie Verslagen*, Vol. I. (1893), pp. 143....

1894 "Lectures on Mathematics." By F. Klein. Evanston Colloquium. Macmillan and Co., N.Y.

"Lines which can be placed on a surface of the third class or third degree." By E. G. *Nouvelles Annales de Mathématiques*, Paris, Vol. XIII. pp. 138....

"On the special form of the general equation of a cubic surface and on a diagram representing the twenty-seven lines on the surface." By H. M. Taylor. *Philos. Trans. Royal Soc.* Vol. CLXXXV. Part I. (A), pp. 37–69.

1898 "Forms of surfaces containing twenty-seven real straight lines." By W. H. Blythe. *Proc. Camb. Philos. Soc.* Vol. IX. pp. 6....

"On the construction of models of cubic surfaces." By W. H. Blythe. *Quart. Journ.* Vol. XXIX. pp. 206–223.

1900 "Ueber die Gruppungen der Doppeltangenten einer ebenen Curve vierter Ordnung." By H. E. Timerding. *Crelle's Journal*, Vol. CXXII. pp. 209–226.

1901 "Die Konfiguration (15_6, 20_3), ihre analytische Darstellung, und ihre Beziehungen zu gewissen algebraischen Flächen." By R. Funck. *Archiv der Math. und Phys.*, Leipzig (3 Reihe), Vol. II. pp. 78–107.

"On models of cubic surfaces." By W. H. Blythe. *Quart. Journ.* Vol. XXXIII. pp. 266–270.

"La configuration formée par les vingt-sept droites d'une surface cubique" (Méthode assez simple pour arriver à leur position mutuelle). By J. de Vries. *Archives Néerlandaises des sciences exactes et naturelles*, Haarlem (Sér. 2), Vol. VI. pp. 148–154.

1902 "To place a 'double-six' in position." By W. H. Blythe. *Quart. Journ.* Vol. XXXIV. No. 1, pp. 73, 74.

1903 "The double-six configuration connected with the cubic surface and a related group of Cremona transformations." By E. Kasner. *Am. Journ. Math.* Vol. XXV. No. 2, pp. 107–122.

"On the Brianchon configuration." By A. Henderson. *Am. Math. Monthly*, Vol. X. pp. 36–41.

"Ueber die Beziehungen zwischen den 27 Geraden auf einer Fläche 3. Ordnung und den 28 Doppeltangenten einer ebenen Kurve 4. Ordnung." By M. Zacharias. Diss. Rostock. Göttingen (Druck v. W. Fr. Koestner).

1904 "On the graphic representation of the projection of two triads of planes into the mystic hexagram." By A. Henderson. *Journ. El. Mitch. Sci. Soc.* Vol. XX. pp. 124–133.

1905 "A memoir on the twenty-seven lines on the cubic surface." By A. Henderson. *Journ. El. Mitch. Sci. Soc.* Part I. Vol. XXI. No. 2, pp. 76–87; Part II. Vol. XXI. No. 3, pp. 120–133.

"Notes on the geometry of cubic surfaces." By W. H. Blythe. *Mess. Math.*, Camb., Vol. XXXIV. pp. 139–141.

"On Models of Cubic Surfaces." By W. H. Blythe. Cambridge University Press.

1908 "On the property of a double-six of lines, and its meaning in hypergeometry." By H. W. Richmond. *Camb. Philos. Proc.* Vol. XIV. pp. 475–477.

1909 "An elementary discussion of Schläfli's double six." By A. C. Dixon. *Quart. Journ.* Vol. XL. pp. 381–384.

1910 "Note on the double six." By A. C. Dixon. *Quart. Journ.* Vol. XLI. pp. 203–209.

"On double-sixes." By W. Burnside. *Camb. Philos. Proc.* Vol. XV. pp. 428–430.

"Notes on the theory of the cubic surface." By H. F. Baker. *Proc. London Math. Soc.* Ser. 2, Vol. IX. Parts II. and III. pp. 145–199.

"On Geiser's method of generating a plane quartic." By (Miss) M. Long. *Proc. London Math. Soc.* Ser. 2, Vol. IX. pp. 205–230.

1911 "A geometrical proof of the theorem of a double six of straight lines." By H. F. Baker. *Proc. Royal Soc.* A, Vol. LXXXIV. pp. 597-602.

"The double six." By G. T. Bennett. *Proc. London Math. Soc.* Ser. 2, Vol. IX. pp. 336–351.

Printed in the United States
By Bookmasters